TORPEDOED

TORPEDOED

An American Businessman's True Story of Secrets, Betrayal, Imprisonment in Russia, and the Battle to Set Him Free

EDMOND D. POPE

and Tom Shachtman

Little, Brown and Company
BOSTON NEW YORK LONDON

For information on Time Warner Trade Publishing's online publishing program, visit www.ipublish.com.

For more information about Edmond Pope's ordeal in Russia, visit his Web site at www.edmondpope.com.

Library of Congress Cataloging-in-Publication Data

Pope, Edmond D.

Torpedoed: an American businessman's true story of secrets, betrayal, imprisonment in Russia, and the battle to set him free / Edmond D. Pope, Tom Shachtman.

p. cm.

Includes index.

ISBN 978-0-316-34873-7 ISBN 0-316-34873-2

1. Pope, Edmond D., 1946– 2. Prisoners — Russia (Federation) — Biography. 3. Businessmen — United States — Biography. 4. Imprisonment — Russia (Federation) 5. Espionage, American — Russia (Federation) I. Shachtman, Tom, 1942– II. Title.

HV9715.15.P67 2001

364.1'31 — dc21

[B] 2001038594

Designed by Stratford Publishing Services, Inc.

Printed in the United States of America

To Cheri

SPECIAL REMEMBRANCE

While none of us will ever forget nor completely heal from the tragic events of September 11, 2001, I would like to pay special tribute to several known victims. On that fateful day, the Chief of Naval Operations Intelligence Plot (CNO IP) was one of the areas struck by terrorists as they flew American Airlines flight 77 into the Pentagon. The Officer-in-Charge plus six other men and women of CNO IP are among the missing and presumed lost:

Commander Dan Shanower
Lieutenant Commander Otis Tolbert
Lieutenant Jonas Panik
Lieutenant (Junior Grade) Darin Pantell
Ms. Angela Houtz
Mr. Brady Howell
Mr. Jerry Moran

Additionally, a friend and former neighbor, Commander Charles "Chuck" Droz, U.S. Navy (retired), was a passenger on that plane.

There but for the grace
of God go we.
May God rest their souls
and protect their families.

Contents

TORPEDOED

CHAPTER 1

Four Days of Hell

The Sayani was nobody's dream of a hotel. From the Soviet era, situated in a far Moscow suburb among mammoth, cookie-cutter blocks of apartments and victory gardens, it was large, impersonal, shabby, out-of-date, and devoid of charm. In a half-dozen stays there, I had never seen another American. During a previous visit, there had been flocks of refugees from Kosovo, then under siege from the NATO bombing campaign. At other times the hotel was overrun with groups of schoolchildren from the hinterlands, in Moscow to gawk at the capital. To see anything of Moscow from the Sayani would require a telescope: there was not even a subway line nearby, though one could find a bus that would take passengers to a subway stop several miles away. But the Sayani was perfect for me: it was on the road to Sheremetyevo Airport, and more importantly, it was near Bolshov's home. Bolshov was my Russian subcontractor, and proximity to him made sense.

It was Monday, April 3, 2000, the finale of a three-week trip, my twenty-seventh to Russia in a decade, and I planned to use it for wrapping up business details and readying myself for the long flight home to the U.S., scheduled for 1:00 P.M. the next day. Russians always want to have just one more meeting to review their positions, make their final offers, monopolize your last few hours, and leave you with a good impression; so, invariably, the last day of one of my stays in Russia was hectic.

To make it a little less crazy, that day I used the Sayani as my base for meetings with various combinations of people from the scientific institute known by its acronym, NIICHIMMASH, with Anatoly Babkin, the Bauman University professor who had been doing one of the research projects that I was there to commission, and others. These meetings also included Bolshov, our translator, and Dan Kiely, my American technical expert from the Applied Research Laboratory at Penn State.

I tried never to meet with Russians in my room, because as far as I was concerned, even though the Soviet Union had been declared dead for a decade and this was democratic, capitalist, post-Communist Russia, doing "business as usual" here meant being continually aware that my room was probably bugged; accordingly, I usually conducted meetings elsewhere in the hotel — the lobby, the tearoom or bar, or, preferably, at Russian institutes or other facilities. That morning I cadged a meeting space in an empty dining room, until it filled with lunchtime diners and we had to move to an awkward spot in a hallway. The meetings were productive — we were discussing deals my company was making with the Russians to commercialize several technologies originally developed for military use — but they were taking too long, principally because of Dan Kiely, who had the tenured-professor's inclination to blather and fuss about details when the rest of us were obviously trying to hurry and reach last-minute decisions.

Still talking, we moved toward 701, my room on the seventh floor, at around 1:30 in the afternoon. The group had dwindled to myself, Bolshov, translator Titov (Bolshov's son-in-law), old Professor Babkin, and Kiely. We were going to the room to collect Babkin's coat, hat, and notebooks. It might be spring in the U.S., but winter lingers in Moscow, and the sixty-nine-year-old professor liked to wear several layers of clothing. I hoped to finish quickly and shoo them out, because I had other business and a personal matter to attend to, dinner with a Russian I'd known for ten years.

After our group had spent a few moments in the room, I decided to shut the door because of noise coming from the hallway. Reaching the door, I was surprised to find about a dozen men standing outside it; the hallway had been empty when we'd come out of the elevator. I started to close the door, but the men prevented me from doing so and pushed into the room.

There were at least a dozen of them, young and middle-aged men in civilian clothes. My pulse was racing, but I remained outwardly calm. I didn't see any guns, but I thought the intruders might be thugs trying to rob us because we were speaking English and therefore could be presumed wealthy. Then I realized that two or three of them carried video cameras that were rolling, and I knew immediately they must be KGB: no media people had any reason to interrupt us in our rooms. Several intruders put their fingers to their lips in the gesture that in almost any language indicates that everybody is to be silent. Kiely and Titov started to speak anyway, and the men grabbed each of them and indicated more sternly that there was to be no discussion.

I didn't understand the insistence on silence, but later guessed it was simply designed to prevent us from screaming for help. When we asked what this intrusion was about, a couple of them flashed identification cards. The Russians and I recognized these cards immediately — you can buy such IDs at certain souvenir shops, and I even had one on my wall at home — but in case we English-speakers didn't realize what was before us, an officer said, in English, "FSB." When Kiely mumbled that he didn't know what that was, the officer shrugged, pointed to the badge, and explained, again in English, "KGB, KGB." Kiely stood bolt upright and became visibly pale and nervous, more so than the rest of us were.

My pulse and mind raced: we would be taken in for questioning, then harassed a bit; I wouldn't be able to meet my friend for dinner, I wouldn't get many last-minute things done, but it wouldn't be too bad. They'd check our documents, grudgingly believe our stories, see that we had no classified material, issue some bogus warning, and we'd be on our way in time to make our plane the next day. When the plane cleared Russian airspace we'd have a good laugh and drink a toast to our adventure.

No guns were drawn, no handcuffs displayed, no expressions of hatred or evil satisfaction creased the faces of the men who'd invaded my room. But their calm, professional, and courteous manner deepened my feeling of alarm. Harassment and attempted shakedowns were part of doing business in Russia, but this, I started to realize, was something different.

The men sat us down on the several chairs and bed in the room and told us — through a translator, and while the videotape was running — that they were going to separate us, and then they "just had a few questions for us." The majority of the KGBers escorted Kiely, Babkin, Titov, and Bolshov out of my room, leaving with me an officer, a translator, a guy with a video camera, and two guards. They shut the door and my questioning began.

The investigative officer was fairly short, relatively young, well dressed, and obviously intelligent. With his trim moustache, slender facial features, and longish hair he reminded me of somebody, but I couldn't figure out who. He asked me who I was, what I was doing in Russia. Each question took a while, because it had to be translated into English, and then my answer translated back into Russian. He wanted to know about, and to have me identify, everything that I had with me, so they could "inventory" it. "Are these all of your belongings?" "Do you have anything hidden? We are going to search your room." "Do you have anything cached elsewhere?" "How long have you been in Russia?" "When are you planning to return home?" I cooperated, reasoning that if I did, they would soon conclude that I was an ordinary businessman doing science-related business.

From time to time, the phone in my room would ring. I was not allowed to answer it, and my questioners also did not answer it.

When I wanted to relieve myself, one guard would stand by the inner bathroom door so that I couldn't make a sudden dash for freedom. They must have thought they had a really bad guy on their hands.

What attracted the most interest from my questioners was a bunch of papers I had been given by the Russians with whom I was working in Novosibirsk, St. Petersburg, and Moscow. These, plus various printed brochures, pamphlets, and the like, amounted to a stack more than a foot high, a typical collection from one of my trips to Russia. About each, the young FSB guy asked, "Where did you obtain this one?" "Who gave it to you?" "What is its purpose?"

My answers to the questions were a bit complicated, because they varied with each document and reflected the fact that I was in Russia working on several different contracts and projects for three different employers — the Applied Research Laboratory (ARL) at Penn State University, a private company, and the Office of Naval Research, which is an arm of the U.S. Navy — as well as for my own companies, CERF Technologies and TechSource Marine Group. I remember saying very clearly that as far as I knew, I had received and was carrying no classified documents, though I did describe at least one document as dealing with a "sensitive technology." After each response that I made in English, the FSB translator would translate what I had said into Russian. The questioning about the documents became more detailed; and after I had explained each one's source and contents, the investigator would put it into an A or a B pile. I quickly surmised (though I did not say so) that the A's interested him, while the B's were of no further interest.

As I had feared, the A pile of documents was inventoried, wrapped in plastic, sealed, and stamped with identifying numbers, and then the investigator told me they would be taking me in for further questioning. That I had seen this coming helped me to remain in control of my emotions. But then I was further alarmed by their next instruction: I was to pack up all of my other belongings and get them ready to be put into the hotel's luggage storage room for safekeeping. When I appeared distressed, the investigator said that my further questioning would only take a few hours, and then I'd probably be released and could come back to the hotel to pick up the bags and go to the airport. That made sense, so I did my packing. There seemed nothing I could do to affect the course of events. I reasoned that I was still in possession of my passport, airplane ticket, and cash, so it was still likely that I would soon be let go.

There followed an odd scene at the desk of the hotel. The clerk didn't want to keep the luggage, but immediately ceased his objections when my escorts said a few words, presumably identifying themselves. Nevertheless, he insisted I pay in advance the charge for luggage storage, which was 32 rubles, about $1.10. I had 30 rubles in my pocket, but none of the almost-worthless single-ruble coins, and to prevent any further delay one of my FSB guards reached in his pocket for the two rubles. I promised to pay him back, and he benevolently waved away the offer and laughed.

I was escorted out to the street and into a waiting black Volga sedan with curtains in the back window, a VIP vehicle of the sort featured in every old novel, movie, or news clip about the Soviet secret police. I was sandwiched into the back seat between two burly guys. For some reason, the car remained at the curb for a while, and during that time I spotted my friend, the man with whom I was expecting to have dinner. He was standing around, a puzzled look on his face as he peered out at the street and into the building. Not wanting to have him recognize me, react, and thereby subject himself to possible questioning by the authorities, I kept quiet in the car and did not acknowledge his presence. I was glad I hadn't written my appointment with him into my Palm Pilot. The Volga soon left the hotel's grounds without my friend having been identified; I took satisfaction from that small victory.

We headed toward Moscow, and I expected to be taken to Lubyanka, the infamous prison and office complex two blocks from the Kremlin that had long been the headquarters and prison of the KGB and its predecessors. In the post-Communist era, Lubyanka still stood out as an icon of terror, the place where at least three chiefs of state security had been executed, along with countless other party officials, functionaries, supposed dissidents, and the like. Lubyanka was a dark landmark.

But we did not go to Lubyanka. The car headed toward another location, and I was told that our destination was Lefortovo. That we were not going to Lubyanka was of some relief to me, but Lefortovo was no less sinister in reputation. Built by Catherine the Great, it was reputed to have niches in the walls of an underground hallway where executioners with silenced pistols concealed themselves before emerging to shoot in the back of the head an enemy of the state being walked along the corridor. Persistent rumors told of beatings and tortures at Lefortovo, but I kept repeating to myself: This is the new Russia, not the old Soviet state. The men taking me in have been pleasant, even courteous. No threats. No raised voices. "Just a few questions and you'll be on your way again."

It was around four o'clock, and beginning to become dark, when we pulled up to Lefortovo. There was some confusion at the outer gate, and I was told the delay was due to the absence of paperwork that was supposed to be there ahead of me. Did that mean the FSB had made the decision to pull me in much earlier, and not, as my polite interrogator had said, at the last minute and because of the materials I was carrying? This was more serious than I had imagined. I might not make that plane tomorrow.

The outer door was open and we went through it; as we did, I had a sense of doom. My escorts flashed their IDs as we passed a guard post, and we were buzzed through an electronically opened grill door. When that inner door clanged shut behind me with an echoing sound redolent of a prison, my stomach did a somersault. Now, I thought, I'm really in for it. They could beat me in here, I could scream, and no one outside would even know it had happened.

The hallway was brightly lit for a Russian building — dark by American standards, though — and reasonably well maintained. I was escorted into a room with a red light above its door, one of many along the corridor. My interrogator from the hotel went in, put his coat into an armoire, and then sat at the desk; it quickly became clear from his body language that this was his office.

The room was eight by fifteen feet, and he sat in one corner, behind a large desk, in a position from which he could see what was happening in the hallway through the door, which was halfway along the interior wall. His desk had a computer on it, the screen toward him. I was put in the diagonally opposite corner, seated behind a much smaller, empty desk whose metal legs were bolted to the floor. Between us, off to the side, an interpreter stood — he would walk between the interrogator and me to make certain that he heard and translated everything. And sitting in one of the several rickety chairs in between the desks was another bureaucrat, who I was told was my "advocate" or "lawyer." A barred window looked out on another wing of offices, perhaps thirty feet away. I became aware of the distinctive decorative feature of the room, a large lithograph directly behind the interrogator's desk, a revolutionarily heroic pose of "Iron Feliks" Dzerzhinsky. Now it struck me: the interrogator had the same moustache, hairstyle, and slender facial lines as Russia's archetypal secret policeman.

Dzerzhinsky was a name and an icon to conjure with; every serious student of Russia — and I had certainly become one over the years — knew of him. Feliks Edmundovich Dzerzhinsky rose to prominence with the Communists, during the Russian Revolution of 1917, as head of the Military Revolutionary Committee of the Petrograd Soviet. After the revolution this group became the Cheka, an acronym for the All-Russian Extraordinary Commission for Combating Counter-Revolution, Speculation, and Sabotage. The word *cheka* in Russian means linchpin, and to every Russian the Cheka came to mean the hated and feared secret police. Encouraged by Lenin, and given the task of smashing opponents of the regime, Dzerzhinsky became a harsh fanatic, ruthless in his use of terror. It is estimated that his Chekists executed a half-million people in ten years. Dzerzhinsky was also the head of the Cheka's successor organizations, the GPU and the OGPU, before his death in 1926. Every Russian entity has to have a hero, and Big Feliks's zealous pursuit of the enemies of the state made him a hero to the employees of the police apparatus that had metamorphosed into the *Komitet Gosudarstvennoy Bezopasnosti,* the committee for state security, the KGB.

I remembered seeing Dzerzhinsky's statue, and its separated base, by the river near Gorky Park in the early 1990s. That seemed fitting, because it had been the KGB generals' attempt at a coup d'état against Gorbachev that had backfired and led directly to the pro-democracy Yeltsin years — during which the KGB had been dismantled and its functions scattered among several other agencies. A few years later, the statue had been righted, but the base was apart. Soon the statue had been reunited with its base, and on a succeeding visit, I found that both base and statue were upright, polished, lovingly restored. And there was talk of transporting it back to its position of honor in front of Lubyanka. The stylized lithograph of Dzerzhinsky above the head of my interrogator at Lefortovo — whom I couldn't help but thinking of as Little Feliks — was evidence that Dzerzhinsky still remained the hero of the state security apparatus.

The interrogation that had begun in my hotel room continued apace at Lefortovo. I had to explain the origin, significance, and background of every paper, in excruciating detail. When my watch registered ten in the evening, and we still had lots of papers to go through, I knew it was going to be a long night, and that the probability of making the morning plane had grown more remote. No dinner was offered, nor did I ask for it. I should have been tired, but I was full of adrenaline and wide awake.

As we progressed, the personnel in the room began to take on individual personalities. The interpreter, Alyosha, who spoke good English and seemed competent but overly friendly, was something of a snake. The man acting as my state-appointed "lawyer," another fairly short guy, in his late twenties, was a mere functionary — not as bright as Little Feliks — and he was certainly no advocate for me.

When I objected to a way of doing things, the lawyer would tell me (through the interpreter), "That's standard procedure. That's normal. Nothing to worry yourself about. Just answer the questions."

I tried to answer questions, believing it was not in my interest to refuse to respond. The training in resisting interrogations that I had undergone as part of my education in the U.S. Naval Intelligence Service helped me focus on what to reveal and what not to reveal. I had nothing to hide about my current activities, but there were many things I didn't want to tell my interrogator. So I told a few white lies and tried to figure out what sorts of information I could conceal or gloss over and what I couldn't. Many facts could be discerned from the materials I carried, but I did not think anything in those materials was illegal.

Some questions concerned the money that, on behalf of my various employers, I had paid to individuals and institutes. Still other questions dealt with whether or not certain materials were classified. I was adamant that none of the materials in my possession were classified, stating repeatedly that because I had been aware that my various contacts dealt in both classified and nonclassified work, I had insisted they *not* share anything classified with me.

"We know a great deal about what you have been doing. We've been watching you," Little Feliks told me. I wasn't prepared to accept at face value that they knew *everything* about me, although I was certain they had indeed been watching: on the first day of this trip, I had come back to my hotel room after a meeting and had noted that the receiver of the phone was a different color from the one that had been there when I checked in, and thereafter had proceeded on the assumption that my phone was tapped. I tried to delicately probe Little Feliks to learn the extent of their knowledge, with minimal success.

Through the open door of the interrogation room I heard something I was not supposed to hear: the voice of Dan Kiely. I had been concerned about Dan, because he was sixty-eight years old, overweight, and — when I'd last seen him — pale, sweating, and agitated. Little Feliks had assured me during the evening that Kiely was "all right," and then I heard Kiely's distinctive high voice, from another interrogation room or from the hall, saying, "Oh, I forgot: There's just one more thing." A few moments later the corridor door clanged, and I thought: They're releasing him — and keeping me! But no one would tell me if that was true or not. My stomach experienced a sudden sinking sensation. What had Kiely told them? What materials might he have had in his possession that could be detrimental to me? Why would they let him go and keep me?

It was nearing three in the morning, after an interrogation that had lasted without significant interruption since dusk the night before. Little Feliks now said that they had too many outstanding issues and would have to hold me for a few days until the questions had all been answered, after which I would be released and could get another plane reservation to go home. With that, a guard escorted me out of the interrogation room.

In a basement room, at three in the morning, I was strip-searched. Nobody probed my cavities — they were careful not to touch me — but the search was an indignity. Everything came off except my wedding ring. After almost thirty years of marriage, my knuckle had expanded so that the ring would not slip over it. They joked about cutting off the ring, or cutting the finger off so that they could make certain the ring was not a secret transmitter. I didn't find this funny. The guards also joked about several items I carried, and their eyes bulged when they counted the $4,700 in dollars that I had with me — the equivalent of four or five years' salary for one of them, I knew. They retained my wallet, keys, watch, the other contents of my pockets, reading glasses, shoelaces, belt, and tie. In another room I was issued a mattress, a single blanket, sheets and a pillow, a bowl, a cup, a spoon, and a fork, and was made to carry them all — shuffling in my unlaced shoes — to detention cell number 68. The door slammed shut behind me; I stood there, trying to wake from this nightmare and unable to do so.

Another eight-by-fifteen room. Three steel tables meant as beds. One sink, one toilet, one small mirror embedded into the concrete, no bars, one opaque window. It was cold in the cell — not freezing, but 45° Fahrenheit, kept at that temperature to make me miserable. I paced for a while. Nervous. Upset. Confused. Unable to sleep. Everything in the world went through my mind: I will get out of here, I won't get out of here, best-case scenario, worst-case scenario. I tried to figure out what they were looking for. Little Feliks had given me an important piece of information: that the paperwork to interrogate me had been approved by the Chief Prosecutor on the 28th of March, a week before they detained me, and while I had been on a visit to St. Petersburg.

I lay down on the mattress, under the blanket, and was so cold that I had to put on my sport coat to keep from shivering. It was dim but not dark, as there was a light on in my cell that never went out. Every few minutes the quality of light coming from outside the opaque glass would change, and I guessed that the guards were checking on me, making sure I hadn't tried to commit suicide or send a message from the Flash Gordon transmitter concealed in my wedding ring. Suicide was the furthest thing from my mind.

As I reviewed the day and night, I recognized that Little Feliks had been playing mind games with me, being nice — "We're your friends. Just tell us everything and you'll be on your way. Would you like a cup of coffee?" — to trick me into revealing more than I wanted to. I was afraid that I had done just that, though I had no idea of precisely what I might have said that could be construed as damaging. I had drunk the coffee, but Little Feliks had not — did that mean they might have put a special ingredient in the coffee that could weaken me? And what was I to discern from Little Feliks's refusal to tell me whether Babkin, Bolshov, or Titov was still being held and questioned? There had been an awful lot of questions to me about Bauman University and Babkin.

There was no sleep that night.

In the morning, I accepted a cup of the coffee — it hadn't hurt me the night before, I reasoned — but did not eat the proffered food. My stomach was upset, but I was also afraid the food was drugged or poisoned. I hadn't done anything wrong, in Russia, but there were matters from my past that needed to remain concealed. I knew that without food it would be difficult to maintain my strength, to resist the interrogation. Nonetheless, I couldn't risk eating.

The second day's interrogations were similar to the first, with curious interruptions. The sessions didn't start until ten in the morning, though at 6:00 A.M. I'd been told to get up, and there was a one-hour lunch break during which I was returned to my cell. I took the initiative only once, asking Little Feliks and Alyosha about an implement in my room. It was a twelve-inch-long, sharpened plastic knife. "If you don't want me to commit suicide, why leave a knife in my cell? I could have slit my wrists." They had no answer, because providing me with a knife to cut food was the rule, similar to the rule that they must provide me with a nonprotecting lawyer, and just as absurd.

I had asked this lawyer to phone my wife, Cheri, and let her know where I was — even gave him the home and work numbers; he said he had called, but I didn't believe him. He also pestered me to sign a contract and begin paying for his services at the rate of $500 per day — an amount equal to about half the annual income of an average Russian. Since he was doing nothing for me, I refused, saying I could not sign an agreement until I had an opportunity to discuss it with someone from the American embassy. That served to deter him from further pursuing the matter. But I wondered: was he trying to tell me that if I paid his exorbitant fee, as a bribe, I would be let go?

Little Feliks's questioning changed to focus on my background. My bio was easily available, so there was no need for me to hide it: I had spent a quarter-century in the Navy, and most of that time in Naval Intelligence. "Aha, intelligence!" Little Feliks observed. The fact that I'd served in intelligence couldn't have been news to them, but they acted as though they'd hit the mother lode. Were they so incompetent as not to have known this about me? Was their reaction just a matter of interrogation style? Another trick to get me to reveal things?

From time to time Little Feliks would leave the room and I would seize the opportunity to ask Alyosha the interpreter what was going on. He wouldn't answer such questions, and when I requested a toothbrush and toiletries, he'd say, "We'll see what we can do." Nothing would be done, and I learned that his responses were simply meant to placate me.

I began to understand the rhythm of the interrogating. I would be questioned for an hour or two at a clip, during which Little Feliks would take notes in longhand. Then he would type up a storm on his computer, frequently calling over Alyosha (and on occasion my attorney) to consult on what he was writing; these were the "protocols," the summary results of a group of questions and answers. I deduced that the computer was part of a LAN, a local area network, which meant that Feliks could access what was being typed in another interrogation room — the one where Dan Kiely had been — and that the material on Feliks's computer could be read and critiqued by his superior.

When a protocol was finished, it would be translated and read to me in English, and I would be asked to agree to what it contained and to sign it. When I objected, my lawyer told me this was the routine, and that we would have serious problems if I refused to sign. Almost invariably, the protocol would misrepresent the questions that had been asked and answered. For example, a protocol would state that I had traveled to a forbidden location to talk to Professor ABC, but it would not mention that my visit had been made at the invitation of ABC. When I raised such an objection, Little Feliks would sometimes amend the protocol, but more often he would simply wave away my protest, telling me that my interpretation of the story would be taken into account later on.

Later on? That was ominous. But I signed the protocols anyway, since refusing to do so would not, I believed, hasten my release.

At two in the morning I was grateful to be returned to my cell. I still couldn't eat or sleep. But at least they hadn't beaten me, tortured me, thrown me into a cell with some violent monster. Visions of electroshock tables, feet burning, testicle burning, teeth-pulling began to recede though not yet to vanish from my mind.

Shivering, I lay awake and mulled over what I'd said and what I hadn't. Little Feliks had informed me that a diplomatic note had been sent to the American embassy, alerting them that I was being held for questioning, so I had to assume that my family had been notified. How soon would I be able to go home? I urged myself toward discipline and control. Why couldn't I sleep? Take your mind off the interrogation, I instructed myself, think about things you enjoy. Think about the family. . . . No, don't think about the family, or you'll upset yourself even more. Think about them anyway: Cheri, the boys, my one-year-old grandchild. Did Cheri tell my dad, dying of bone cancer, that I was detained in Russia? Would that news accelerate his decline? Think about a nice walk in the Oregon woods where you grew up, a raft trip on the Rogue River. . . . Maybe you'll never be able to experience such things again.

On the third day, April 5, I received a visit from the American embassy. I had little advance notice before being summoned from my cell and escorted to a room slightly larger but not much different from an interrogation area, where I met my visitors under the supervision of an FSB guard.

Brad Johnson of the embassy staff was middle-aged, personable, and courteous. He was accompanied by the embassy's own translator, Igor, who helped me ask questions of the FSB. I wanted to know what had happened to Dan Kiely. The chief interrogator said Kiely had been released, but he wouldn't tell me whether or not Dan had gone back to the U.S. "Did he get all of his belongings back?" "I can't tell you that," was the FSB's translated response.

You are a private citizen in a foreign country, subject to local laws, Johnson told me. Here is a list of local lawyers; we can't help you choose among them because of State Department regulations.

Without additional information on the ability, inclination, and reliability of those on the list who worked on criminal cases, I could have thrown a dart at the names, and in a sense I did, picking out one, whom Brad said he'd call for me. Meanwhile, he wanted me to be certain that I emphasized in my defense the distinction between industrial-secret gathering and espionage. I told him I was not guilty of either. He simply nodded. Since we made no headway on that, I raised other subjects. Yes, your wife has been notified that you are here; yes, we can visit you once a month but not more often; yes, we can bring you food and other things if we've received money to pay for the supplies and the FSB will let them in. As he went through this litany my level of annoyance rose, because he clearly seemed to be implying that I was guilty of whatever the Russians planned to charge me with — and this, while I was sitting there, telling him I was innocent. Yet I had to accept what Johnson said at face value, because I had no other source of information about what the U.S. government could do for me, or what I could do for myself.

Near the end of the third day, the interrogation halted and the guards hustled me to another area of the complex for what they said were "routine matters." I was given a razor and a hairbrush — neither had been provided to me before — and told to shave and clean up. After I made myself presentable I was photographed, mug-shot style, fingerprinted, and taken to a "clinic" room for elementary measurements: height, weight, blood pressure, pulse. I'm sure my pulse was racing, because these routine matters were evidence that the bastards were not going to let me go in a day or two — else they would not have bothered processing me in such a manner. That made me very nervous.

Still, they might bring charges against me, I reasoned when alone in my cell, but then they would let me go, expel me, and forbid me to return. That would be bad, but it wouldn't be awful. Their insistence on cleaning me up for photos proved that this whole thing was political, was about seizing an American for the propaganda value they could squeeze from the situation.

Friday, the fourth day, dawned, and I had still had no sleep. More interrogation, more protocols, a sharper focus on the materials dealing with the Shkval torpedo, whose propulsion system we had been pursuing for use in civilian applications. Questions from previous days were repeated, word for word. More accusatory statements: "You were seeking restricted information."

This was incorrect, I explained, as I had already done many times: "We had a contract with Bauman to supply us with *authorized* information."

There were more reductions to the protocols, couched in terms damaging to me. For instance, on another project we had gone by invitation to a semisecret facility several hours north of Moscow; but when we had arrived, our clearance had not yet come, so rather than ask the officials to skip that procedure, we'd said we'd go back to Moscow and wait for them to call us when they had obtained the required papers. I told this to Little Feliks, but when he finished processing the information, his protocol read something like, "Americans went to semisecret city, were refused admission, and returned to Moscow without getting what they wanted." I objected to a formulation that left out my full explanation; all that remained was a sentence which implied nefarious behavior on our part.

"Putting in your details is not important," Little Feliks said.

"Yes, it is important!" I responded with some heat. By this point I could no longer hide my agitation.

Little Feliks, a fairly mild man, refused to shout me down. "If this process goes any further," he'd say calmly at such moments, "you will have occasion to explain everything in your own words. This is just the protocol. You can always challenge it."

At the end of the fourth morning's interrogation session, I was escorted back to my cell for lunch. After an hour, when I was returned once more to the interrogation room, things had changed. Alyosha told me that matters were too complex to be settled in a few days, and that they would be holding and questioning me for a longer period. Significantly, for the first time, he didn't mention anything about my going home. Instead, he said, "We're going to put you in a cell with others like you."

"Others like me? Am I a leper?"

"No, you're a spy."

"A spy!?"

"Well, that's what we think you are."

Of course, from four days of questioning I had understood that they were heading in this direction, but because during that time they had never made the actual accusation, I had entertained a false hope that they would not charge me with espionage. Similarly, no one prior to this moment had used the words "court," "arrest," or "charges." It was devastating news.

To regain balance, I asked again about being put into a cell with "others." What others?

"Spies, terrorists, serious criminals," Alyosha ticked off.

Little Feliks jumped back into the conversation, telling me that this phase of the interrogation was finished, and that he wouldn't be seeing me anymore. On Monday, he said, a new interrogation would begin, with a new interrogator, one he characterized as — in contrast to himself — "a tough guy."

For the next few hours they let me stew about this in my isolation cell. I became steadily more agitated. It was a mind game, I knew, to leave me for the weekend with the hint of menace that a tough guy would be taking over, to let me know that I had had it easy so far — but I couldn't help permuting the terrible possibilities. I believed I had been resolute. I hadn't broken down, hadn't lost my will to live or to resist, despite four nights without sleep and having undergone lengthy interrogation. When would they start the torture? How much longer could I keep myself together? What sort of monsters would they lock me in with? I had not known until then how much I'd valued being alone in a cell.

Around four in the afternoon the guards came. They spoke no English, and only with difficulty conveyed that I must gather all my belongings — mattress, bowl, etc. — and carry them. They wouldn't touch any of it, wouldn't help me. We walked down an interminable series of halls, until we came to cell number 80, whose door was closed and shuttered, obscuring what was within. I had no idea of what evil might await me behind this sinister-appearing portal.

My heart pounded as the guards opened the door and half-ushered, half-pushed me into the cell and shut the door behind me. The cell was physically much the same as my earlier one, with the three steel bed-foundations, toilet, washbowl, small mirror. A young Russian stood there, quite normal looking, tall and thin, nonthreatening, and evidently a bit surprised to see me.

He stuck out his hand, and intoned his name. "Sasha."

"Ed," I said, shaking his hand.

He looked at me askance. Ed is not a name regularly used in Russia. There are Edmunds, but not Eds. Almost no Russian name consists of one syllable. Just like that, I had let him know that I was not a Russian, as he had assumed, but a foreigner, an American.

CHAPTER 2

Behind the Green Door

There were reasons why the Russians might consider me a spy. Chief among them, I had done intelligence work for over twenty years in the United States Navy. So, imprisoned in Lefortovo, not knowing what would happen, whether my family knew where I was, or whether the government I had served for a quarter-century would help me, I looked backward for a clue that might explain what had just happened.

My involvement with the military began while I was in college in the mid-1960s. The conflict in Vietnam was escalating rapidly, and friends were returning from tours of duty there with wounds, severe drug and alcohol habits, and changed mental attitudes just a whisker away from being deranged. I didn't want any of that happening to me, and while I had no intention of shirking my duty to serve in the military, I wanted to find a better way to do so than as a rifleman in a ground platoon. So at Oregon State University during the spring of my freshman year, 1966, I joined the NROTC, the Naval Reserve Officers Training Corps: that would keep me from being drafted so long as I stayed in school (itself a helpful spur to my lagging academic attention), pay me a small stipend, and assure that when I did go on active duty I would be better prepared and trained, and in a safer job.

A typical Oregon State student, I hailed from a small town and had a middle-class background and a Protestant Anglo-Saxon family tree that went back in the U.S. for at least four or five generations. My father, Roy Edmond Pope, was an engineer by calling, though he had not finished high school during the Depression. He and my mother, Elizabeth, both originally from Oklahoma, met and married when they were working in different capacities for the shipyards in San Diego during World War II. I was born in San Diego on August 4, 1946. My parents moved to oil-rig country in Texas when I was quite small, and my sister and brother were born there. In 1954 the family visited a relative in Oregon for two weeks, decided we loved the place, with its cooler climate, deep forests, wonderful fishing streams, comforts, and opportunities, and moved to Grants Pass, where my father purchased a small logging-equipment business. My dad proceeded to expand the equipment company and become a prominent local businessman. Other than the tragedy of my brother being killed by an automobile in an accident, in this almost idyllic setting our lives were pleasant, relaxed, and enjoyable. My father's one business blossomed into three. My mother did her part in those businesses, raised us, and after finishing nursing school, worked as a nurse for the next twenty years.

In high school I was a middling student who enjoyed mathematics, ran long-distance races, and had a moderate talent for tinkering that echoed my father's greater ability to manipulate and fix all sorts of mechanical devices. Oregon State, 150 miles away, had a good math department. In the summers, I took a job with the U.S. Forest Service, helping survey the wilderness for roads: it was out of doors, healthy, and invigorating, and it utilized my training in math and provided tangible evidence that the work I was doing was practical — in contrast to most of my courses at college, which seemed less and less relevant to the real world. As a student I did well only in courses that interested me, principally those dealing with computers, which were then clunky machines that took up whole rooms.

In an era when many students in the United States of America were in full-throated revolt against authority and in opposition to American participation in the conflict in South Vietnam, I held different views. I loved my country and thought it had few serious faults. I was in favor of stopping the spread of communism and in doing so in Vietnam. I also viewed the protests as a strength of our country and believed in protecting freedom of speech — so long as the demonstrations did not inflict harm. But I understood, as few of my contemporaries did, how constrained our servicemen were in attempting to defeat the Communists in Southeast Asia. My NROTC classes — and friends who served in Vietnam — taught me about the many restraints placed on the troops by politicians.

I graduated in 1969, just before the summer of Woodstock, the manned landing on the moon, the Haight-Ashbury district of San Francisco being overwhelmed by pot-smoking hippies. That fall I joined the Navy as a reserve officer, an ensign. I had no intention of doing more than fulfilling my three-year obligation and perhaps going on to become a computer programmer. I was sent to San Diego for specialized training — in computers and as an air-traffic controller for shipboard-based aircraft, specialties that combined my skills of mathematics and mechanical prowess. During the training period I met Cheri Thompson, who had just applied to United Airlines to become a stewardess. In May 1970 we were married and immediately thereafter had to go in separate directions: I shipped out on the U.S.S. *Oriskany,* an aircraft carrier that served as a platform for testing new computer equipment and that was going to the waters off Vietnam, while Cheri went to Chicago to attend United's stewardess-training school.

I hated my position on the *Oriskany,* an ant in a huge colony of 3,500 people, one of many very junior officers serving under a frustrated boss whom we found to be contemptible. The air-traffic control hours were challenging and rewarding, and so were those spent on the advanced computer systems, but most of my days were taken up with make-work, hurry-up-and-wait, and the other bugaboos that sailors and soldiers usually complain about. We were in a combat zone but were never directly shot at. We did lose several pilots during the cruise, as well as a crew member who fell overboard and was never found, and the ship's doctor, who got caught in the middle of a gunfight in a small Philippine city: these experiences sobered us by bringing the reality of the conflict to our immediate attention.

During our first venture into the Gulf of Tonkin, as an air-traffic controller I accidentally vectored two fighter aircraft over Hainan Island; fortunately, they exited Chinese airspace before there was any reaction from China.

The *Oriskany*'s stops in Hong Kong, the Philippines, and Japan were highlights of my seven-months' duty and gave me a love of travel and visiting exotic places that has never slackened. The ship was on its way back to San Francisco in late 1970 when orders came to move me to Hawaii and into intelligence work. It was highly unusual for a reserve officer to be transferred to such an interesting place and position, but Cheri and I saw it as an opportunity, and she left the airline (and a job she liked) to set up house near Pearl Harbor. Years later, I would realize that my move to Hawaii was part of the Navy's beefing up its intelligence staff in the Pacific to support the conflict effort in Vietnam.

In 1971, then, I went behind the green door — intelligence spaces in the Navy all have green doors and no windows — to join a group of two dozen, which grew eventually to fifty or sixty, handling a new worldwide, automated, computerized, information-processing system. Our purview was the whole of the Pacific and Indian Oceans. We monitored and analyzed coded information from surface ships and submarines, uncoded commercial-ship radio broadcasts, material from satellites, electronic surveillance, human observers, and other sources. The conflict in Vietnam was still going on, and our superiors were also concerned with such matters as the war between India and Pakistan. Now a lieutenant junior grade, I was exposed to the interplay of geopolitical forces, the need to keep continual track of friendly and unfriendly naval assets. I was also exposed to a fascinating world of secure telephones and Teletypes connected to Washington and to other Naval Intelligence centers around the world, with 24-hour-a-day vigilance, access by special clearance only. I learned that my intelligence-officer peers were smart, aggressive, and self-motivated; in civilian life they had been engineers, mathematicians, lawyers, and journalists, and were considered some of the best of the junior officer corps. Being continuously challenged intellectually suited me just fine. Already cleared for "confidential" and "secret" matters, while I was in Hawaii the Navy conducted a thorough background check on me prior to awarding me Top Secret and Special Intelligence (SI) compartmented clearance: friends, neighbors, employers, and teachers had to vouch for my integrity and trustworthiness.

The information collected and analyzed behind the green door at Pacific Fleet headquarters was focused on the task of supporting operational forces. It had to stay behind our door unless and until we took it to the operations side, for example a report of the delivery to North Vietnam of missile-armed KOMAR-class Soviet-built patrol boats, an incident that could involve one of our submarines, news of Soviet bombers leaving a base and heading toward one of our aircraft carriers, or a report of a Soviet sub that had showed up unexpectedly in one of our sensitive operating areas: all these were the sort of matters that the ops people needed to know about immediately.

From collecting and carrying information, after a year I stepped up to the more difficult task of analyzing it. What were the patterns, the past history that could enable us properly to assess data and to predict activities that might have an impact on our Navy's forces? We couldn't be too alarmist in our evaluation, but equally we couldn't afford to ignore what might be significant. I learned that the secret of being a good intelligence officer lay in the judicious choice of adverbs such as "probably," "possibly," and the rarely used "certainly." When I was unable to figure out the meaning of a new piece of information, I'd get on the Teletype to someone more experienced. There were many such instances. For me, a milestone was reached when other analysts started coming to me for answers to their questions.

In the early 1970s we began to notice a quantum leap ahead in the quality of Soviet naval forces, and a commensurate expansion in their scope and in the sophistication of their operations. The reason was then unknown; today, in retrospect, many of the advances and unusual activities have been attributed to the result of John Walker's espionage against the United States. Starting in 1968, Walker, his family, and his associates stole the keys to encrypted communications and delivered them to the KGB in what the Navy later labeled as "the most costly act of espionage in the history of our country." The Walker spy ring was not shut down until 1985. It is now believed that the Walkers' treason enabled the Soviets to learn — to their chagrin and, in some instances, to their probable shock — that the U.S. was far more advanced than the U.S.S.R. in certain maritime technologies; in reaction, it is believed, the Soviets commissioned research institutes to produce new and countertechnologies, whose results began to show up in the equipment of Soviet naval forces after a compressed development cycle. Other American turncoats, most notably Ron Pelton of the National Security Agency (NSA), by their actions compounded American losses.

At about the same time that I was promoted to analyst, the entire Pacific Fleet intelligence division in Hawaii came under the command of Captain Bobby Ray Inman, a studious, visionary man already marked by Navy scuttlebutt as clearly headed for the top.

Intelligence analysis was exciting and full of responsibility. For example, one of my tasks was to monitor North Korean naval developments. The North Koreans had recently begun a modernization effort that included buying and building guided missile boats and submarines, both of which were potential threats to U.S. Navy assets. My predecessor in this job had been removed from it, and his future in the Navy placed in doubt, for failing to warn the U.S.S. *Pueblo* that the threat level from the North Koreans was extremely high. (However, he was one of the officers who did forward a last-minute alert that the *Pueblo* was about to be detained and boarded, in what became a very serious international incident.)

My job had only one drawback: I couldn't go home and tell Cheri much about it. I might say to her I'd had a good or a bad day, an exciting or a dull one, but not much else. Occasionally I'd be able to advise her to watch for something special to show up on the evening television news broadcast. I'd often have to do double shifts and work at night, but the work itself was downright enjoyable. I spent every spare minute reading about weapons systems, strategy, electronics, and politics; so did my cohorts who wanted to advance.

The time was drawing near for me to either leave the service after my required three years or extend my active duty. Cheri was doing secretarial work in Honolulu part-time, and caring for one-year-old Brett. My work was fascinating, a promotion to full lieutenant was in the offing, we enjoyed the beaches and our new friends, and we decided that I should extend my tour for a single year, though still as a reserve officer, and see how things went.

Paradise was very seductive. That year flew by, and intelligence work and the life in Hawaii were still so attractive that I shifted over into the regular Navy and we bought a house. Now a "1630," a career Naval Intelligence officer, I was sent on short trips to Japan, South Korea, and other countries, where I interacted with counterparts in the navies of friendly forces. Captain Inman gave me a few special projects to do; he was testing me, though at the time I didn't know what for. Six months later I found out, when my detailer in Washington told me that Inman, who had preceded me there to become Rear Admiral Inman, Director of Naval Intelligence (DNI), wanted me to join the Naval Intelligence staff in the capital.

It was early 1975; Nixon had been replaced by Ford, South Vietnam was crumbling rapidly, there was a great need for intelligence about the shifting patterns in the world, and I was a twenty-eight-year-old lieutenant in the Naval Intelligence Service who had already been personally tested and approved by the director of that service. Admiral Inman wasn't finished with my education yet, either. After I'd spent a few months in an outlying office, he assigned me directly to the Pentagon. To give me personal exposure to important people, both in uniform and in the offices of the civilian secretaries of the Navy and of Defense, and to allow me to obtain a broader view of the Navy as a whole and its interactions in the world, I was to become a briefer, working under Commander Tom Brooks.

As I soon learned, this was a swim-or-sink job.

The Chief of Naval Operations (CNO) Command Center, which is open and humming with people and information-flow 24 hours a day, 365 days a year, has three parts: the Command Center itself, which deals with the status of American naval forces everywhere in the world; the Intelligence Plot, where Naval Intelligence personnel use all sorts of sources to locate hostile and potentially hostile military forces in the world; and the CNO Briefing Theater, which resembles the best-equipped movie theater in the world, with state-of-the-art electronic display equipment and individual executive lounge chairs.

Every morning at 7:30, the Navy would conduct a three-part briefing for two or three dozen flag officers and other big shots in the one-hundred-seat theater on the fourth floor of the D ring corridor at the Pentagon. First up, each morning, would be the Public Affairs officers, speaking about the headlines of the day; then Operations would report on U.S. Navy ship positions, casualties, and the like; then came Intelligence. For our briefings, we'd have to ask certain people to leave the room, since they weren't cleared for the information we would present. The people in the Intelligence Plot area had worked all night, following late-breaking events, preparing the material, and rehearsing the brief, but for an individual briefer like me, it was still frightening to stand up and address these three- and four-star admirals, all of whom had decades more experience than I did, and some of whom had deserved reputations for asking very tough questions. Young briefers could be and often were chewed up if they had not properly prepared themselves and weighed the meaning and context of every word they uttered; before we gave the actual briefings, Tom Brooks would first "murder board" our proposed presentation to prevent us from self-destructing.

We delivered our brief, and thereafter, in the theater, spoke only when spoken to. When we couldn't answer a question, Brooks would try to do so. During the rest of the day we'd conduct research or brief individuals or small groups on specific matters; I did one such briefing for Les Aspin, then chairman of the House of Representatives' Armed Services Committee. Rich Haver, a senior civilian analyst, went with me; Aspin's office had to be swept ahead of time to ensure that there were no listening devices.

Typical of the special items about which we would prepare reports for the morning briefings were frequent updates on the operation of the U.S.S. *Batfish* in 1978; for seventy-seven days, the American attack submarine conducted a trailing operation of a Soviet Yankee-class ballistic missile submarine. Even though the Soviet submarine was engaging in strategic patrol operations against the United States during this time period, the *Batfish* managed to remain undetected as it trailed the enemy sub.*

We also prepared a daily magazine, highly classified, called the *CNO Intelligence Briefing Notes*, for distribution to a short list of people in the Pentagon, other government organizations, and in key naval units around the world. As I steadily became cleared for more and more projects, I was expected to keep current on them so that I could be prepared at any moment to brief or write about such matters as the seizing of the *Mayaguez* by Cambodia, the sinking of a Soviet submarine, or the launching of a new class of Soviet subs, etc. There was a lot of pressure on a briefer, but it was a plum job, the equivalent of being a deck officer on a rough, seagoing command. Several of my predecessors in the Intelligence Plot had made admiral, and one still serves as a U.S. senator; on the other hand, two of them made the front page of the *Washington Post* after being arrested during a demonstration protesting the conflict in Vietnam.

*This operation is described in an article in the March 2001 issue of *Smithsonian* magazine.

Hard work in the Navy is not enough; what job you hold and which people in positions of authority are able to see the good work you do also count when you are being considered for the next-highest rank and for your next posting. My personnel file had a tab indicating that Admiral Inman would make or approve any change in my job or location; he had similar tabs on the files of fifty to a hundred other officers, people he knew and trusted. I didn't mind having a star watch over me, especially one who knew the ways of the Navy and commanded the respect of all who knew him. One of the other tenets of the Navy is that you can't stay behind a desk too long and expect to make admiral. In 1978, Inman decided that I needed a two-year, seagoing, operational tour of duty to continue to be a "hot-runner" marked for continued promotion and challenging tasks.

I reported to the staff of the Sixth Fleet, aboard the U.S.S. *Albany,* a medium guided missile cruiser, the only one of the Sixth Fleet's three dozen ships to be permanently based in the Mediterranean, at a small city named Gaeta, on Italy's west coast between Naples and Rome. After the Vietnam Conflict, the Sixth Fleet was where the action was: Middle East tensions, NATO exercises. The Navy paid for my family to relocate, and we rented the upper part of a villa. Our second son, Dustin, was born in Italy.

The American and Soviet navies in the Mediterranean would keep track of one another, but with different styles. One day, Admiral James Watkins and I were in a helicopter, photographing Soviet ships for fun rather than as serious intelligence-gathering. He wanted to see them up close, and I was able to identify the various types of vessels, their weapons systems, and their purposes, as well as take a few snapshots. When we neared a Soviet vessel, its decks would always be bare of people and no equipment would appear to be operating. In contrast, whenever a Soviet aircraft or ship would approach the *Albany* in the Med — we'd know it was coming, from our intelligence sources — the crew would make it a point to stand on deck and give friendly waves, and to put brightly colored boxes and strange-shaped appendages on various protuberances, just to confuse their observers.

On the *Albany,* I continued to brief every morning; and I was under the command of a more senior Inman protégé, Bill Studeman, who would also go on to become the Director of Naval Intelligence, a full four-star admiral, and Deputy Director of the CIA. One of the tasks of the *Albany* staff was to be ambassadors; one month the ship would visit Israel, the next, for political balance, Egypt. This was pleasant duty, with great opportunity for sight-seeing and for learning some history. Frequently, Cheri would fly to wherever we docked and join me for such excursions. Our scrapbooks filled up rapidly.

Three incidents stand out from my two years on the *Albany*. The first involved an official visit to Monte Carlo, during which Prince Rainier came aboard the *Albany* and Admiral Watkins and I engaged him in a private discussion. We sat in the admiral's ward room and briefed the prince; the Admiral spoke about the status of American and friendly forces in the Med, and then I spoke about what Soviet and other presumed unfriendly naval forces were doing in the region. The prince was most attentive and courteous, and remained with us well beyond the time allotted for the briefing.

The second incident involved America's then-simmering dispute with Mu'ammar Al-Qaddafi, Libya's leader. Qaddafi, who had come to power in 1969, in 1977 had declared Libya to be a socialist Arab republic and was increasingly aligning his country with the U.S.S.R. He was also voicing brash threats and belligerent statements regarding Libya's claim to vast areas of what were actually international waters in the Mediterranean. In an exercise designed to convincingly demonstrate the right of any nation to operate freely at sea — and to show Qaddafi a bit of the naval power that we could bring to bear on Libya, should he make any aggressive moves — the *Albany* and other Sixth Fleet ships parked right off the coast of Libya, yet well outside Libyan territorial waters, and fired some missiles. Even though these missiles could travel at speeds up to three times the speed of sound, or Mach 3, they were considered out-of-date ordnance that the Navy was trying to use up, and they served as target practice for planes from one of our carriers; the planes would attempt to intercept the launched missiles and shoot them down over the Gulf of Sidra, the least-populated portion of the Mediterranean, which just happened to be near Qaddafi's shoreline. This exercise could easily have escalated into an international incident had anything gone wrong, but nothing did. Qaddafi, who had adopted the habit of sleeping in his tent in a different part of the desert every night, got the message.

The last event of note involved a trip to Romania. Annually, an American surface ship would sail into the Black Sea, just to have a look around, to show the flag, and, again, to demonstrate the right of freedom of the seas. I did not travel there on the *Albany* but aboard a different ship, on which I was billed as an assistant to the admiral in charge. Actually, I had been promoted to lieutenant commander and was on the trip to observe and report. Romania was the first Communist country I had ever visited, and the visit was an eye-opener. Dictator Nicolae Ceausescu had been in power in Romania since 1965, and although he made a show of independence from Moscow, Romania's ties with the Soviet Union were very close. Taking photographs only in permitted areas, I nonetheless made a small intelligence coup, confirming for the first time that the Soviets had sold advanced MIG-23s to the Romanians.

Ashore, the plight of a Communist country was palpable. In the port city of Constanta, as well as in Bucharest, we saw hard-currency stores with well-stocked shelves, catering only to foreigners; these were in contrast to stores that sold only to locals with Romanian currency — stores that had virtually nothing to sell, and customers who had no money in their pockets. People were arrested and beaten on the street right in front of us, some for trying to approach our ship, others for attempting to exchange local currency for dollars. We were taken to visit an "orphanage," really a boarding school for children taken away from their parents to be indoctrinated. Our hosts also took us on a ride to a winery. They were proud of their infrastructure in Romania, as exemplified by a divided, four-lane highway. But the highway was empty of cars in both directions, and in several sections battalions of women with brooms were sweeping the concrete. No wonder the Romanians could boast of full employment.

In early 1980 we went back to the United States for a year, in preparation for my going to Sweden as assistant naval attaché, a post Inman had once held. Part of my training was language and culture school, and part was knife-and-fork school — table-setting, etiquette, seating of dignitaries, protocol. Cheri was encouraged to participate in those schools, since she would be going with me. We both learned passable Swedish. At language classes Cheri and I met Dave Moss, a naval aviator who was going to be the senior naval attaché in Sweden, and his wife, Joan; we became fast friends.

Wives were excluded from several weeks of training given to us at a secret facility by the Department of Defense, the CIA, the FBI, and other government agencies. We learned to resist interrogation, avoid hostile encounters, evaluate potential terrorist threats, drive defensively, and the like. My cohorts at the course were men who were slated to be attachés, as well as those from the State Department and from other government agencies who were shortly to be stationed overseas in other capacities. Among our guest lecturers were American diplomats and military personnel who had been in difficult situations and escaped, and Soviet defectors, who taught us what to expect from Soviets in overseas postings. It was not just classroom instruction. We played roles in mock interrogations and at mock cocktail parties. At the cocktail parties, when we were the good guys, our objective was to find out pieces of information A, B, and C from other guests; at the same time, those playing the bad guys were instructed to obtain information tidbits D, E, and F from us. At the mock interrogations, the goals were similar. After such interrogations, the tapes would be replayed for us, and we'd be critiqued on how well we had accomplished our tasks; we'd also learn techniques to better avoid revealing information, and to determine what the interrogator actually knew as opposed to what he implied that he knew. The course was at times lighthearted but the underlying intent was deadly serious. Within months after these secret sessions, an Army officer from our attaché class was assassinated on the streets of Paris by a terrorist. That sobered us all.

There was a requirement that all Navy and Marine Corps officers heading out of the country for attaché duty had to meet with Admiral Hyman Rickover, the "father" of the American nuclear submarine fleet. Rickover took the opportunity to tell a bunch of us that what we would be doing in such duty was "wasting the taxpayer's money," and to threaten us by saying that if any of us betrayed any secrets to the Soviets, even inadvertently, he would visit his well-known wrath upon us and wreck our careers. Having belittled and warned us, Rickover then stormed out of the meeting room.

Before Dave Moss and I went to Sweden, we met Captain Lennert Forsman, then the Swedish naval attaché in Washington, and when we arrived in Stockholm we were delighted to find that Forsman had returned home and was now the commander of the Karlskrona Naval Base. We made plans to visit him — an involved process requiring clearances, one month's advance notice, and particular dates of arrival and departure. While waiting for the papers, in Stockholm we encountered the Soviet naval attaché, Yuri Prosvernin, a cold, rather paranoid Siberian who our information said was the GRU — Russian military intelligence — man at the Soviet embassy.* In the Cold War days of the first Reagan administration, neutral Sweden's reputation as a haven for spies seemed well deserved.

On October 27, 1981, Dave and I drove the 300 miles to Karlskrona, arriving in town in the evening, and staying overnight at a hotel before going to see our friend the next morning. We were in Lennert's office no more than fifteen minutes when his deputy rushed in, very agitated, and whispered in his ear. Lennert had to

*The Soviet ambassador to Sweden was Evgeny Primakov, known to be a KGB officer; he would later become prime minister of Russia under Yeltsin.

leave the room, and for about an hour we sat there, hearing people hurrying excitedly about, until Lennert told us what had happened. About the time we had arrived in town the previous evening, in the fog and drizzle a Soviet submarine had run aground on the rocks near the base, and first light had revealed its location. It was what NATO called a Whiskey-class submarine, an old diesel; news reports around the world instantly dubbed the incident "Whiskey on the Rocks." Years later, several books would be written about this incident, and one of them even claimed that Dave and I had planted an electronic device along the Swedish coast that had affected the Soviet submarine's navigation equipment in a way that resulted in the grounding.

The entire country of Sweden became alarmed at the clear violation of its neutrality that the grounded sub presented. At first the Soviets would not permit the Swedish Navy to board the submarine, and of course the Swedes then refused to assist in pulling it off the rocks. By evening, the normally quiet town of Karlskrona had become a beehive of military and government officials, besieged by news reporters. Two days later, Yuri Prosvernin arrived in Karlskrona and was incensed to see Dave Moss and me; he accused us of meddling, of trying to create an international incident out of what the Soviets claimed was a navigational accident. There was no way to convince him that our presence in Karlskrona was a coincidence. The Soviets threatened to make a formal diplomatic protest to the Swedes about our continued presence, so we told Lennert that if he wished, we would leave immediately.

Someone else also wanted us to leave: the American chargé d'affaires, a career diplomat and the number two man at the embassy in Stockholm, who called us at our hotel in Karlskrona and told us to get out of there before we created an international incident. Fortunately, cooler heads in the Department of Defense prevailed, and we were told we could stay if the Swedes wanted us to. And Lennert did; he argued that the trip schedule had called for us to remain there three days, so we must adhere to that schedule. We stayed, but were circumspect in our behavior.

When the Swedish government was allowed to board the submarine, the boarders found, among other instances of possible tampering, that the ink on the log entries of the four days before the grounding was hardly dry.

The Swedish Navy insisted on checking the sub's manifest and thus learned that there was a senior flag officer on board, which was highly unusual. His presence, and certain other matters involving the crew and the submarine's equipment, indicated that the sub had likely been on an espionage mission, trying to get information on Sweden's antisubmarine-warfare munitions, being tested nearby, and/or to map a recently realigned minefield.

Behind the scenes — because to advertise our presence could have caused more complications — we told Lennert that the sub was likely to be carrying nuclear-tipped torpedoes. Our posture was, We don't want to tell you how to do your business, but if it were our problem, this is what we'd try: point sensitive detectors at the section of the sub where torpedoes were usually stored. The Swedes had already done so, and they revealed to us that the detectors' needles had gone off the charts, identifying the presence of the nuclear-tipped torpedoes. They were concerned about potential radiation leaks from the submarine, even though Yuri denied that possibility because the sub was diesel-powered.

What happened next exemplified the bind that the Swedish authorities believed they were in as a nonaligned, determinedly neutral country in close proximity to the Soviet Union. For a week the Swedes tried to discuss privately with the Soviets the presence of the nuclear warheads in Swedish territorial waters. Only after receiving a healthy dose of Soviet arrogance, denials, and refusals to cooperate did the Swedes let reporters know about the nuclear warheads. The resulting outcry in Sweden was enormous, and thereafter the Swedish public began to look upon the Soviet Union with much less benevolence. "Naval attaché" Prosvernin had to eat his words, suffering public embarrassment. The incident led to Sweden's setting off explosives in its coastal waters to flush out, or at least discourage, Soviet subs from further espionage.

Dave and I further angered Yuri Prosvernin at a cocktail party by teasingly asking him to sign a book about the KGB* that listed him as an agent — he refused — and succeeded in making him foam at the mouth at a memorable attachés-only skiing trip above the Arctic Circle. All the military attachés stationed to Sweden went on expeditions together once or twice a year. On this one, unable to ski properly, Yuri hit a rock and broke off one of his Swedish-issued skis; I retrieved the broken tip and wouldn't give it up when he demanded its return — it is mounted and framed in my study today, next to an aerial photo of the grounded submarine in the "Whiskey on the Rocks" incident. In the six months after the incident, Yuri disappeared from the diplomatic circuit three times, and on each occasion, when we inquired about his absence, his Soviet colleagues told us he had gone back to Moscow because his mother had died. On the third such occasion, we considered sending flowers. The dossiers of Dave Moss and Ed Pope at the Soviet embassy became quite full.

When I was detained by the FSB in Moscow in April 2000, Dave and other friends recalled our tweaking of Yuri Prosvernin and thought he might have had me arrested as revenge for my actions in Sweden many years earlier. Two items suggest otherwise: (1) Prosvernin was GRU, not KGB, and the two agencies were more rivals than colleagues; (2) if Yuri had had a hand in my arrest and detention, his personality was such that he would have made certain I found out it was his doing — and no hint of that ever surfaced.

* * *

*Actually, it listed him as an agent of the GRU. *KGB: The Secret Work of Soviet Secret Agents,* by John Barron. Reader's Digest Press, 1974.

My three years in Sweden passed very quickly. We dealt with Soviet defectors, European security conferences held in Stockholm, high-ranking congressional and State Department delegations, and the like. I learned that the embassy's State Department personnel viewed themselves more as advocates within the American government for the point of view of the country in which they were serving than as employees of our government whose job was to advocate the American point of view to the host country. I should have paid more attention than I did to that particular lesson, but, involved in my own work, I did not fully absorb its implications.

On my return to the United States, I visited Admiral Inman in his new post — deputy director of the CIA. We chatted a bit in Swedish. During my years abroad, Admiral Inman had also served as head of the National Security Agency. At the CIA he reached the level of four-star admiral, the first man from Naval Intelligence ever to achieve that rank. Soon afterward, Inman retired and took a position in private industry.

At the Pentagon, to which I returned as a commander, I was given an unusual one-year assignment as executive assistant to Rich Haver, a senior civilian analyst in the office of the DNI. Haver was another man on his way to the top; today he is a senior assistant to Secretary of Defense Donald Rumsfeld. My official title was head of the Soviet Strategy and Tactics Branch — but the branch was only three people, all working in support of Haver on special, highly sensitive projects. In this new position I received so many clearances, for our various ultrasensitive projects, that it was difficult to keep track of them. Some projects were so secret that fewer than twenty people inside the Beltway were cleared for them. The projects included supervision of ultrasensitive intelligence gathering and research and development of weapons and other devices. Even today I am not permitted to discuss the details of these programs, or even to offer an opinion as to whether what is printed about them is true or false.

I was the junior member and executive secretary of a working group of top brass and analysts like Haver and the most senior decision makers in the Department of the Navy. We would regularly arrange reviews and discussions for these senior people about advances in intelligence and weapons development. It was like being continually engaged in a futuristic war game. For instance, the ability to use smart bombs, electronic jamming and deception systems, and rapid and accurate surveillance for bomb damage assessment could change the tactics for countering particular kinds of threats — and we'd take that into account in our calculation of how much danger a potential enemy posed to our national security. At times we participated in actual war game exercises at the Naval War College and other places; we'd have to carefully sanitize our information even for people with advanced clearances, because often we knew of things that those playing the games did not even dream were on the drawing boards and might be available in a future war. Advising a chief in a war game, I'd tell him, in effect, "Shoot the guy in the foot, not in the head." The chief would not need to know what I knew from intelligence sources, that the foot was the enemy's weak spot. He simply had to have confidence that my directive would enable him to prevail.

There was so much secret information in my head just then that I was not permitted to leave the country without written permission, and certain areas of the world were ruled off-limits to me. Had I been captured at that time, and forced to reveal what I knew about our intelligence and future weapons developments, the present and future security of the United States could have been seriously compromised.

A member of the working group, Mel Paisley, then the assistant secretary of the Navy for research, engineering and systems, requested that I next work directly for him. It was to be a short assignment, after which I had lined up a two-year sea tour that would help me keep in the running to have a shot at making admiral. (One of the ways that people of promise are recognized within the Navy is by their series of short assignments; if you are serving a full three years in a slot, you are probably not slated for rapid advancement.) Paisley's office controlled an R&D budget of $12–$15 billion and involved the research of more than forty thousand people across the country, inside the military and in contractors.

Because of a change in administrations, Paisley resigned not long after I had begun working for him. He was later convicted for his earlier activities in that position, in connection with the procurement scandal known as Ill Wind, and spent several years in prison. He later emerged from prison a changed man and began a new career, outside the bounds of the areas that had originally brought him to his position in the Department of the Navy, one which brought him new and positive distinction.

Until there could be a new political appointment to the assistant secretary position, the job was filled on an acting basis by Dick Rumpf, who had risen through the ranks of the Navy technical community. Dick was probably one of the most effective people ever to serve in a temporary assignment of this sort. Shortly thereafter, Thomas F. Faught Jr. was politically appointed as the assistant secretary. Tom liked to emphasize that he and I had a lot in common, being graduates of Oregon State and working in intelligence — he in Iran, before the fall of the shah.

I had been in the job about six months, well before Paisley's departure, when a routine medical examination revealed a tumor on the ribs on my left side. At Bethesda Naval Hospital the growth was removed and certified as cancerous. But the hospital's laboratory couldn't identify the tumor and sent it to the National Institutes of Health, which distributed parts of it to various labs around the country for analysis. The consensus of the analysts was that my disease was a rare form of bone cancer called hemangiopericytoma. Not much was known about it, but the data suggested that if you could survive the first two years without a recurrence of tumors, you could probably go on to live a normal life — but for those first two years you had to be closely monitored. Stress was also identified as a major factor in the cancer's recurrence.

For me, to be diagnosed at age thirty-nine with this kind of cancer and prognosis was a double whammy: it placed me in danger of imminent death, and it was a career-ending illness. I knew immediately that I would never again have a post at sea and would never make admiral; the Navy would do its best to keep me healthy and would take care of me, but because of my medical condition it would never put me in a higher position of command or authority.

I'm not the sort of person who gives up when faced with adversity. I still stood a chance of making captain, and had the opportunity to do some good work. My background in mathematics and mechanics had given me a lifelong appetite for R&D, and assisting Rumpf and Faught — and moving within the spooky world of intelligence — I was happy and challenged. The only person on the staff of two hundred who had the same clearances as Rumpf or Faught, I would often be designated to substitute for them in meetings, and to follow developments while they paid closer attention to the big picture and administrative matters. Among the programs I helped manage was a series of meetings in major cities with defense contractors and research institutions to critique what in our R&D programs the Navy was doing well and what needed revision.

I also helped supervise the Navy's efforts to create unmanned, ultralight, high-altitude aircraft. On their huge wings, such planes can glide for days while using very little fuel, and they are terrific platforms for sensing apparatus whose data is sent to ships or to the ground by high-bandwidth data links. Another project of mine was the development of just such a high-capacity, secure data link over which reams of sensing data could be transmitted. Some of the work involved in creating this data link would become instrumental in the development of the high-capacity switching and transmission networks that supported the emergence of the Internet.

I remained in the assistant secretary's office four years, and the cancer did not recur. In 1989, as President Reagan ended his term in office and George H. Bush was inaugurated, it was time for me to take another assignment. Because of my medical history — and because I was still enjoined from leaving the United States — I did not qualify for every job. I agreed to become commander of the Navy's Pentagon intelligence briefers. The instigator behind the move was Rear Admiral Tom Brooks, my former boss as a briefer, who was now the DNI and needed someone who understood the task.

Once more I would be in the CNO Operations Command Center, this time not as a lieutenant but in charge of the Intelligence Plot and Briefing Theater. I knew this would be a tough job, with long hours and the burden of having a secure phone in our bedroom for the times when I had to be awakened with urgent news from somewhere in the world; everyone who has held the job has come out of it with scars; but doing the job well would almost guarantee that I would make captain.

It was two years of trial by fire, more intense than Brooks or I had imagined possible. The Berlin Wall fell. The Soviet Union began to disintegrate, more rapidly than analysts had predicted, bringing on perilous potential consequences for U.S. military planning.* General Manuel Noriega was extracted from Panama by American troops. And we had Desert Shield and then Desert Storm in the Persian Gulf. I was at the heart of the process and present at the moments when the top Navy brass received and debated the information on which they based their decisions. For me, as well as for all the officers and enlisted ranks who worked for me, this became a time of grand excitement that reinforced our pride at being in the Navy and able to serve our country.

Fortunately, I arrived at the briefing center before the worst of the heat began — because what I found, in the computer systems supporting this most sensitive intelligence operation, appalled me: 1975-vintage Wang computers so antiquated that parts could no longer be obtained. I twisted some arms and cadged a $300,000 appropriation

*The best guess of the Sovietologists, after the fall of the Berlin Wall, was that it would take a decade for the Soviet Union to crumble. It actually took less than two years.

scheduled for other projects, and we managed to rip out the Wangs and substitute an advanced Macintosh LAN network before the United States commenced operations in the Persian Gulf. The new system was all-secure and, under fire, worth its weight in gold. Today I shudder at what might have happened had the Navy been forced to go through the Gulf War with a seriously out-of-date computer system in the CNO Intelligence Plot and Briefing Theater.

At the same time, we were watching the collapse of the Soviet Union and its vassal states. We were very concerned about the Soviet military, a trapped beast all the more dangerous because it was cornered and frightened. Economic constraints had reduced the Soviet military's resources, political unrest was making it nervous, yet it still had millions of men under arms and thousands of nuclear warheads ready to be launched. We received information as power groups formed and clashed, as the trickle of defectors and émigrés widened to a torrent. We worried that the commander of a rogue ship or submarine might take matters into his own hands; this is the scenario depicted in the novel *The Hunt for Red October*, but for us, such dangers had nothing to do with entertainment.

Several times in previous years, Saddam Hussein had massed forces on Kuwait's border, then retreated, so when the Iraqi invasion actually took place it was something of a surprise to the United States, and there was a lot of finger-pointing at people in the intelligence community who should have known this or should have told us that. Once the invasion began, it was near round-the-clock duty for me and my augmented team of briefers and watch standers. We functioned as a situation room, updating information as changes arrived by the minute. Navy ships and installations were attacked by Iraqi missiles and bombs. There were Navy casualties to report, as well as launches of our own missiles from Navy vessels, and bomb damage assessments. We had to hurry up the implementation of new data systems that gathered and analyzed information from satellite imagery in order to provide direct and rapid support to the operational forces in the Gulf.

It was a very involving time for me: endless cups of coffee, no inclination or time to sleep, information pouring in from everywhere, urgent questions. Where is that ship now? What kind of "incoming" is being launched? What does this movement of Iraqi troops portend? Was that facility knocked out, or is the report just pilot exaggeration? Why did Iraq fire a SCUD missile at Israel, one with a cement warhead? I was at the nerve center of the Navy at the moment that nerve center was most active, and I loved it. Although I had no problem being at that post seven days a week, I frequently had to order my people to go home or they would have stayed and worked until they fell over from exhaustion.

For me, the saddest part of the Gulf War was the decimation of the Iraqis; hundreds of thousands were killed as Saddam Hussein sent his troops directly against massed Western alliance forces. The first days of our attacks utilized devices such as smart bombs and special electronic equipment to knock out command and control centers; suspected nuclear, biological, and chemical weapons facilities; radar installations; and missile launch centers. Hussein had deliberately built nurseries, hospitals, and other civilian facilities on top of and surrounding these militarily important sites, knowing full well that attacks on them would produce huge numbers of civilian casualties. Only days after the initial hostilities in the Gulf War, Hussein ordered his troops to store front-line combat aircraft near the bases of ancient sculptures, knowing that the United States and its allies would not risk bombing these artifacts, even to take out the fighter aircraft.

Iraq was a major purchaser of Soviet military equipment, and of course we kept track of how that equipment performed (or didn't perform) during the hostilities. The Soviet Union also seemed to be behind Iran's initial sending of a phalanx of its Soviet-built fighter planes to Iraq to help make war against the U.S.-led coalition of countries. This was a surprise, because Iraq and Iran had been deadly enemies for a decade, Iraq using poison gas on Iran, for instance. Before long, the Iranian fighter planes, and some of Iraq's few remaining aircraft, flew back to Iran to escape our smart bombs and other ordnance, which were destroying them on the ground and inside hardened bunkers. This move, too, was likely orchestrated by the Soviets.

I commanded the briefers well enough so that when my year group became eligible for promotion to captain, I made the grade. (Many others didn't: of the naval officers class of 1969, which had originally numbered about one hundred and had already been winnowed to thirty-five at the rank of commander, only fifteen made captain; the rest, many of them good officers equally deserving of promotion, would spend the remainder of their Navy careers as commanders.) Knowing I would never again go to sea, and never make admiral, I now sought a less physically demanding berth, a place where I could put in three years as a captain before retiring, but that would also be exciting and would enhance what I might do later as a civilian. I found the perfect combination of my skills, knowledge base, and enthusiasms by moving to the Office of Naval Research (ONR).

My predecessors at ONR had generally functioned as security officers for the outfit, examining classified documents before they were turned over to researchers. I had no intention of limiting myself to that role, and as soon as I reached ONR in the spring of 1991, opportunities to do what I really wanted — work to further research and development — sprang up immediately.

The Soviet Union's economy was collapsing so completely that ONR was receiving faxes and letters from individuals and institutes in the USSR, offering to do R&D work for the United States. Defectors would show up on our doorstep with similar offers. Soviet scientists and technologists knew that ONR underwrote not only military projects but also technologies having civilian and commercial potential, and that the ONR annual budget was in the billions of dollars. Since I was the intelligence and security officer at ONR, their offers were routed to me for investigation. Work on these unsolicited offers meshed with my attempts to solve a problem brought to me by friends in the naval intelligence community: what to do with defectors and émigrés who were so brilliant and technically proficient that it took experts in their specialized field to properly debrief them. We set up a program to bring together teams consisting of an American expert, a translator, and an émigré or defector, to write papers about technologies in which the Soviets were more advanced than the West. For such work, we established strict guidelines: we had no interest in intelligence activities or weapons systems development; we were focused entirely on maintenance reduction and systems reliability, and on civilian applications. The qualifications of most of these émigrés and defectors impressed us, as did the quality of the work they had done in the teams, so we found jobs for some of them in private industry; this activity eventually evolved into the Science Opportunities Program, which I created and headed for ONR.

By late 1992 the period during which I could not travel to Russia because of my clearances had come to an end. Émigrés and defectors wanted us to meet former colleagues who had remained in what had been Soviet territory, and there were the senders of the faxed research offers to look up as well, so it seemed logical for me, as a representative of ONR, to visit the FSU's many scientific and technical institutes to determine if alliances and joint research projects could be established. Having spent years on American military-sponsored technological research, and also being somewhat expert on its counterparts in the Soviet Union, I could evaluate what the East Bloc institutes had to offer.

One of the earliest results of the Science Opportunities Program utilized a small group of Ukrainians who had developed a technology called Electron Beam Physical Vapor Deposition (EBPVD). It uses electron beams to coat a metal or ceramic surface with a thin film that prevents corrosion. This is especially important for turbine engines, whose blades can get so pitted by particulate matter that they break up or fail. This was a boon for power generators and aircraft jet engines. Another project proposed by a Russian group was a magneto-hydrodynamic drive similar to the "caterpillar" drive depicted in Tom Clancy's *The Hunt for Red October*. The Russians showed me their equipment and described their progress, but we did not have the technical interest or available funding to pursue the matter.

During the Cold War era, Soviet scientists worked under severe restrictions. An intellectual limitation was that they were not able to travel freely to the West and directly soak up information available to Western scientists, although they could and did obtain some information through the efforts of KGB agents. Conversely, they could frequently obtain money for research that by Western standards was high-risk if they could convince their superiors it would eventually produce something of military value. Lacking supercomputers and the new technologies derived from their advent, Soviet scientists were forced to do a lot of "lateral thinking" and to fall back on areas in which they had excelled for decades, such as theoretical mathematics and physics. The results were evident in such fields as materials technology, production technology, and processing techniques. For instance, the Soviets developed a system for measuring problem spots in the radar grid at airports, not by radar-calibration equipment — an expensive and sophisticated way to do it — but through mathematical modeling, which cost much less. Other important areas of Soviet innovation included aerodynamics, microwave generation, optical sensing and processing, lasers, and propulsion science. The Soviets had perfected a pilot ejection seat that prevented injury to pilots even when they had to bail out at speeds greater than Mach 1, and a WIG, a wing-in-ground-effect vehicle that could fly just a few feet above water or ground, using a propulsion system that pushes air under the vehicle to keep it aloft.

In the three years just after the collapse of the Soviet Union, when I visited Russia under the auspices of ONR, I was welcomed everywhere. The former Soviet scientists there were earning an average of $40 a month, well below what taxi drivers and blue-collar workers could make. I did no recruiting on these trips; even then, we recognized that part of our effort must be directed at preventing any further "brain drain," on the grounds that removing the very best minds of the country would contribute to making the country even more dangerous and unstable. And in fact there was a lot of work for Russian scientists to do where they were, in cooperation with the United States. For instance, during a trip in May of 1993 I met with scientists and technologists from the leading institutes in the fields of oceanology, nuclear reactors, atmospheric optics, materials development, titanium manufacturing, remote ocean sensing, shipbuilding, and marine engineering. I brought back proposals on more than three hundred projects; the sixteen most interesting ranged from new methods for measuring information in the sea to the manufacture of the titanium used for nuclear submarine hulls, to WIG vehicles, techniques for rapid crystal growth, and to a "nonintrusive radiometer" for measuring body temperature. One project involved a liquid polymer that could be sprayed or rolled onto a surface to absorb and remove radioactive spills — it had been in use by the Soviet military since 1975 but had only recently been approved for commercial sale. Another was copper-alloy monocrystals that could be shaped into implements that had "pseudo-elasticity," meaning they were able to be bent or compressed greatly, yet when opened would reextend to their maximum length. These crystals also had "shape memory," meaning that they could be molded, then crushed, and when heated to a certain temperature would return to their original shape. Hundreds of potential uses were suggested for these crystals, from shape-retaining eyeglasses to medical implements that could be inserted into the body in compressed form and would then extend in order to do their work. Russians were perfecting nonmagnetic steel, explosive techniques for welding and bonding, glass-and-plastic propellers that were lightweight and wear resistant but of high strength — a myriad of interesting technologies. The first American ever to visit many of these institutes and previously off-limits cities, I was shown, and was solicited to buy, technologies that a few years earlier Westerners would have been shot for investigating.

What with the usual slowness of U.S. congressionally funded development cycles, I was only able to get a few of these projects up and running by the time I retired as a captain in 1994, but by then I knew I wanted to continue what I'd begun: joint Russian-American development in technologies that the Soviets had pioneered or perfected, and for which civilian applications could now be developed. The Ukrainian-led electron beam vapor deposition project was now sited at Penn State's Applied Research Laboratory, a facility we at ONR had chosen because ARL had many of its own experts in advanced materials. In addition, ARL received the vast majority of its other funding from the Navy, so it was a place we were comfortable with. During my final months at ONR, ARL director Ray Hettche asked me to come to work with him when I retired. I would have a mandate to continue my frequent trips to Russia, now to identify other projects in Russian institutes that could mean new business for ARL. Cheri was offered a job at the university, and we would also get a tuition break for her and for my younger son, who would soon graduate from high school and wanted to attend Penn State. It was perfect.

Thus, after twenty-five years in the Navy I became a civilian again. But I still had in my brain the details of secrets that I had been cleared for over the years I had spent in Naval Intelligence — 126 special clearances on matters of high importance to the security of the United States. You don't just scrub those from your memory. And I did not consider these secrets a burden. Like every other officer who had been cleared for such matters, I expected and accepted that I would go to my grave before divulging any of these secrets, because I understood their pivotal relationship to the national security of my country. They had been entrusted to me by the United States, and I was pledged to reciprocate that trust by not talking about the secret matters. Ever.

Six years later, when the Russian FSB detained me in Moscow and began serious questioning, it was those secrets that concerned me most. I had nothing to hide about my private-sector work, but as the second, "tough guy" interrogation began, I certainly had details from my twenty years of intelligence work behind the green door that I definitely wanted to keep submerged — for my own good, and for the good of the United States of America.

CHAPTER 3

Interrogation and Incarceration

On April 4, 2000, when I did not call from New York to say that the plane had arrived from Russia and that I would be home soon, Cheri naturally became upset. A shy and private person who often describes herself as an ordinary wife and mother, concerned mainly with family, Cheri was not accustomed to major crises, although she had endured plenty of lesser ones during the twenty-five years that she had spent as a Navy wife. Not comfortable or practiced in taking aggressive actions in a crisis, Cheri had nonetheless recently developed new strengths and skills while working for Penn State during a two-year assignment to put together the annual session of the National Governors Association, to be held that year in July at State College, and for which she'd had to make lots and lots of phone calls to people in positions of authority, people whom she hadn't known. So when I did not return from Russia, Cheri realized that something was wrong and that she would have to do something about it.

After a night of very little sleep, Cheri went in early to her office and called Ed Liszka at ARL, the deputy head of the organization, who served not only as Dan Kiely's superior but as my contract administrator. At about 7:30 A.M., she reached a secretary in Liszka's office and asked, "Is Ed there?" A grim charade followed. When the secretary said yes, Cheri assumed she had meant me, and she raced from her office through a rain shower to the ARL office, four blocks away, only to learn that there had been a miscommunication. The Ed who was there was Liszka. But now Cheri was in his office, so she asked him if he knew what had happened to Dan and me.

Liszka told her four astonishing things: that Dan had returned from Russia the night before, that I had been detained by the Russian security services for questioning, that ARL didn't know what I had been doing in Russia, and that ARL could not help her. Cheri knew that I had been under contract to ARL for most of my work on that trip, and in addition, that Dan Kiely had been with me the majority of the time. As she left Liszka's office, Cheri told him that if ARL would not help, she would find someone else who would.

Even more distressing to Cheri would be a statement put out shortly by Penn State's director of public relations, Bill Mahon, that the university had no plans to work for my release. "There is no reason for Penn State to be involved," Mahon told a reporter from the daily newspaper in State College, the *Centre Daily Times*.* At about the same time, someone from Penn State's main administration office called the office of Congressman John Peterson, who represented the Fifth District of Pennsylvania, and informed them that Peterson ought to say nothing about Ed Pope to anyone, certainly not to the press. Peterson remembers thinking that this was odd, because at that point he did not know who I was.†

Kiely remembers an additional odd thing about the behavior of Penn State and ARL. When he returned to the U.S., he phoned the laboratory and was told that they already knew what had happened to him and to me. He assumed that if ARL knew, they would have called Cheri, so he did not do that until the next day, when he and Cheri had a brief conversation.

*Erin Wengerd, "Report: Former PSU researcher jailed abroad." *Centre Daily Times*, April 7, 2000, p. 1A.

†Keith McClellan's handwritten notes of a conversation with Peterson's office on April 7, 2001, confirmed the accuracy of John Peterson's memory of this strange telephone call.

When Cheri returned to her own office on April 5, her supervisors were sympathetic and encouraged her to do what she could to help me. She phoned the State Department, reaching the Russia desk and a bureaucrat named Bill Daniels. He confirmed that I was being held at Lefortovo and lamely apologized for not having gotten around to phoning her. He brushed aside Cheri's mention of my quarter-century in the Navy as reason to assist me, said that I was in Russia as a private citizen, and since I was no longer a government employee there was not much they could do — or, he implied, would do — for me. As though reading from a script, he informed Cheri that State would not help choose a lawyer, though they'd provide a list of them, and that they would only visit me monthly, because they lacked adequate personnel. Cheri asked how many Americans were imprisoned in Russia and was told there were four, but that the embassy's 2,000-person staff was too busy to see them more than monthly. When Cheri voiced incredulity at this schedule, Daniels accused her of being hostile.

A later call to Daniels produced such an eloquent and on-target exchange that it must be quoted. Daniels reported to Cheri that on Brad Johnson's second visit to Lefortovo, I had made three requests: for a blanket, some toilet paper, and a Bible. Cheri asked whether these things had now been provided.

"No," Daniels said, and gave no further explanation, nor offered any hope that the requests would be honored in the near future.

"If my husband asked for a blanket, it's because he's cold," Cheri said. "If he asked for toilet paper, it's because he's sick. And if he asked for a Bible, it's because he's scared. Won't you even try to help him?"

Locked up in Lefortovo, I attempted to understand what was happening to me and could not shake the feeling that it was hugely unjust, because for nearly a decade I had been a champion of Russia's scientists and technologists. More than that, I had become a Russophile: I admired the Russian people, liked many things Russian, would bring back hundreds of souvenirs from every trip, hung Russian art and crafts pieces until they all but overwhelmed the walls of my home. In my garage sat thousands of *matrioshka* — the stacking dolls — that I'd commissioned from Russian craftsmen and was giving away to friends or selling at modest prices. When people told me that friendship between Russians and Americans could never exist, I'd insist that it could, and should, and that my vision of a scientific and technological partnership between the U.S. and Russia would help both countries in the new millennium.

Now here I was in a Russian jail, accused of espionage and mortified at being the cause of my family's suffering. I hoped that Cheri and my friends were working to get me out, but the FSB's isolation techniques kept me in ignorance of any such efforts. My focus during the second week of April was my new interrogator.

Number Two was indeed different from Little Feliks. His lithograph of Dzerzhinsky showed Iron Feliks in a slightly more sinister pose, and in his office it hung in back of my desk rather than behind his own. Vasily Vladirovich Petukhov was tough-looking; a photo of himself that he showed me revealed that earlier in life he had been athletic, but in the interim between his teens and his mid-thirties he had put on considerable weight, and most of it had gone to his posterior (or, as we say in the Navy, to his stern). His most pronounced physical attribute so characterized his mental attitude that I dubbed him Blubber-Butt. Petukhov/Blubber-Butt could compress a half-hour's worth of interrogation into a two- or three-hour session. When he didn't like my answers to certain questions, he would chain-smoke, scowl, shout, and pound the table, but when I was actually being evasive — say, about my knowledge of classified matters or my active-duty assignments — he'd accept my simplistic explanations without challenge. Frequently he would stop the interrogation and quietly contemplate my most recent answer, his brow furrowed in a manner designed to convey the depth and subtlety of his thinking.

As Little Feliks had warned, Blubber-Butt was uncompromising; but he wasn't as bright as he thought he was — or, for that matter, as tough as he believed himself to be. To intimidate me he kept a large FSB officer's cap atop the armoire, but I learned that it, and many other symbolic appurtenances, were just for show. For a few weeks he pretended not to know any English, but then started playing his favorite Frank Sinatra cassette and let on that he knew a few words of Sinatra's and my native tongue. He'd also turn on the radio in the office and pointedly turn it off when the news was announced, in order to keep me isolated from the outside world. However, and as he must have known, the double cell I had begun to share with five other men had a television and a radio from which I obtained quite a bit of news.

Blubber-Butt's loaded questions made it clear that the FSB thought they had caught a major American spy. He accused me of leading a many-tentacled espionage network that included contacts and clandestine operations at the various research institutes that I had been visiting in Russia for years, of corrupting Academicians and other leading scientists with my rich cash pay-offs, and of receiving from them the sort of military secrets that could undermine the security and safety of the Russian Federation. He tried to link me with Russian citizens already under arrest for espionage, such as Igor Sutyagin, who had worked on environmental matters for the U.S.A.–Canada Institute. The FSB was certain that I had microdots, secret transmitters, and James Bond–type spy devices concealed in my toothpaste or within my Palm Pilot.

These absurd charges of espionage were soon repeated by the Russian media, without challenge. Some excerpts from a typical article in *Izvestiya* give the flavor:

> In 1995 Pope retired with the rank of captain (this is equivalent to the rank of colonel in the Russian Special Services). Supposedly he did so at his own request. But at the FSB they believe that the only reason he did this was to stop attracting heightened attention. . . . [In Russia] he presented himself at times as

> a scientist, and at times as a businessman, and the entire time stayed in different hotels and led a fairly modest life . . . in a perfectly professional [espionage] manner. . . .
>
> No one at the Lubyanka had the slightest doubt of what Pope's real goals were any longer. The spy was actively offering cooperation, promised help in introducing technologies in the West, and hinted at investments and major profits. . . . First he bought from his "partner" technologies that were not secret and information that could easily be found in the open press. But he paid generously for it, thereby getting his hook deeper and deeper into the source.
>
> In 1998–1999 contacts between them became regular, and both of them were being monitored by the FSB almost every minute. It was decided to take [arrest] Pope and his supplier of secrets after it became clear that the latter had in fact agreed to pass information to the spy that was really a state secret.

Pretty soon, echoes of the FSB propaganda and the Russian press's repetition of it were appearing in the Western media, where many stories stated, as though it were a fact, that I was a spy. The Russian press also accused me of stealing such things as Pustevoit's acousto-optical tunable filter — which I had legitimately bought and exported, with all the proper permissions — and of nefarious dealings in cities that ranged from St. Petersburg and Moscow to Novosibirsk and Kaluga.

My family and friends denied that I was a spy and pointed out that my years in the intelligence service had been spent as an analyst and a manager, not as an agent. Nonetheless, when Cheri made calls for help, for instance to Pennsylvania senator Rick Santorum's office, the senator's legislative aide stated as though it were common knowledge that I had been an agent for the CIA. When Cheri told him that was not true and asked where he had formed such an idea, the aide told her it had come from press articles. When Cheri similarly called the office of Vice President Al Gore, his staffers refused to assist her in any way and referred her to the Immigration and Naturalization Service, an agency that would have been no more helpful to her than the Department of Agriculture.

When my former colleagues in the Naval Intelligence community learned of my arrest, they at least had an idea of what to do. Before mobilizing for action, they needed to settle for themselves the question of whether I had been a spy. Retired admirals Tom Brooks and Sumner (Shap) Shapiro, both of them former DNIs, and others on the executive board of the Naval Intelligence Professionals (NIP) organization began inquiries. To answer their question — had I been spying? — was not a simple matter, for as I've indicated above, the biggest concern of the intelligence community lay with my 126 clearances, and which of these I might be revealing to the Russians. Some people in the community wondered if I might have told the Russians such secrets in exchange for the many "favors" the Russians had done me over the years: in effect, these Americans were accusing me of spying not on the Russians but *for* the Russians! Off at the other end of the spectrum were some people in the CIA who expressed annoyance at the notion that if I had been a spy, I hadn't been sufficiently under their control: in their view, by not spying I had missed a great opportunity! But more rational sentiments prevailed. Brooks's call to the current DNI, Paul Lowell, obtained the information that I was not employed by Naval Intelligence in any way. Calls to other agencies, such as the DIA and CIA, brought similar answers: Pope wasn't working for us.

Paul Lowell made sure that the Chief of Naval Operations knew that I was not employed by the Navy as a spy, so that the Navy as an entity could — as it shortly did — respond to inquiries from the press, the White House, the State Department, and several congressmen, by insisting that I had no connection with espionage efforts and had been in Russia solely as a private businessman. Rear Admiral Paul Gaffney, then head of ONR, also pitched in to help and to inform all callers that my post-Navy work for ONR was being done by me as a private citizen.

Having settled that matter to their satisfaction, the NIP executive board decided to help me as best they could. They divided up the work among several people: one to handle the press, another to marshal congressional support, a third to assist my family, a fourth to work with the Navy, the White House, and the State Department. Among the NIP-ers who began to work extensively on my behalf were old buddies like Dave McMunn, but also people with whom I had had only served briefly, like Sid Wood, and those with whom I had had no prior personal relationship, like Ted Daywalt, who had been on active duty in Great Britain when I had been an attaché in Sweden. Later, these men would all tell me they felt that what happened to me in Russia could as easily have happened to them, and that they had been motivated in part by the sentiment, "There but for the grace of God go I."

As Cheri received calls from friends asking how they could help, she told them to pray for me. The prayer circles begun by this straightforward request would eventually grow to include a great many people across the country, and the knowledge that many people, even complete strangers, were praying for the Pope family comforted and assisted Cheri — and, when I heard of their efforts, me, too. Also of inestimable help to Cheri were neighbors like Tom and Dotty Pelick, in whose home she would frequently unwind after frustrating days. Our entire neighborhood in State College banded together and was willing to do whatever was necessary or possible to express that support.

Another important call Cheri received that first morning was from one of my partners, State College resident Keith McClellan, who had been alerted by Brad Mooney and had also found information on the Internet about the "American businessman" arrested in Moscow. This was a difficult conversation for Cheri, because she did not know Keith very well and had been somewhat uneasy about the fast friendship he and I had developed. Indeed, our decision to team up (we named our venture TechSource Marine) was relatively recent. But Keith responded to Cheri's confusion and consternation by jumping into action without a moment's hesitation, and thereafter, for the duration of my detention, he worked side by side with Cheri and served as her chief of staff. To begin, Keith — an attorney by training — canvassed acquaintances in Washington for names of reputable Moscow lawyers and on obtaining some, tried to get in touch with them. After receiving a few turndowns from Russian criminal lawyers who wanted nothing to do with an American accused of espionage, he and Cheri reached a prominent attorney named Gofshtien. He was on a case in Siberia, but he recommended his father, Michael Gofshtien, a partner in the same firm, along with a younger lawyer, Dmitri Barannikov. American and Russian advisers had both cautioned against retaining lawyers who were Jewish, because that would give the FSB one more reason to dislike me; but many of the prominent and reputable lawyers available to take cases involving foreigners were Jewish. Cheri and Keith wired some money as a retainer to Gofshtien and Barannikov, and they agreed to start the process of defending me.

This was fortunate, because when Brad Johnson of the American embassy in Moscow had called the lawyer that I'd picked almost at random from his list, that lawyer had refused the case, and Brad wouldn't have occasion to tell me of the turndown until his next monthly visit. At that point, had I been completely dependent on the embassy, I would have had to make another throw of the dart and wait an additional month to learn if my toss had found a lawyer who would agree to take my case. At that rate of progress, it could have taken me a year to find representation.

Another important call that Cheri made during my first few days of incarceration was to Congressman John Peterson. Peterson, after a career as a small businessman, had been a state legislator for nineteen years until he was elected in 1996 to the House of Representatives as a middle-of-the-road Republican. The Popes were not personally known to Peterson, and there had been that mysterious call to him from the administration at Penn State; nonetheless, his office immediately began to work with Cheri and Keith. Peterson himself had seldom been abroad, and as he later told me, he felt unqualified for the task of getting me out of Russia, so he turned for help to another member of the Pennsylvania delegation, fellow Republican Curt Weldon, the acknowledged congressional expert on Russia — the only member of either house who speaks Russian and has extensive contacts in the Duma, the Russian Academy of Sciences, and throughout the Russian Federation.

Peterson's office, too, made a round of calls to the Navy, to the CIA, and to other agencies and was told in no uncertain terms that I was not a spy for the U.S. government. The congressman was happy to learn this; however, he had already determined to help me no matter what the government agencies told him, because, as he would say on many later occasions, "Pope is an American citizen held in Russia, and he needs our help."

Peterson and Weldon began quietly to ask questions of the State Department and to press the Clinton administration to focus on my case and get me out of jail. A third representative willing to actively help in this cause was Congressman Greg Walden of Oregon, who was quite supportive and comforting to my parents at home in Grants Pass, especially to my terminally ill father.

My champions told State and the White House that because Ed Pope stood accused of espionage, which implied that I had worked for the U.S. government to steal the state secrets of Russia, the charge was as much against our government as it was against me personally, and therefore they must do their utmost to refute it and to bring me home. Seated by chance at a dinner next to former National Security Adviser Brent Scowcroft, Peterson spent an hour asking questions and receiving advice on how to proceed, from a man who had been through many international crises. Scowcroft told Peterson that pressure could get me out of Russia, but that dealing with the Russians must involve straight talk and no mixed messages. He also said that the affair would probably take many months before reaching a satisfactory conclusion.

Cheri informed anyone who would listen of her greatest fear: that my rare form of cancer might recur. Although it had been in remission for many years, it had to be watched and treated regularly, and stress was a known factor in its recurrence. Also, since the cancer was so rare, it was unlikely that Russian doctors had enough knowledge to detect and treat it properly.

During this period, the strategy most strongly recommended to Cheri, Keith, and my extended family and friends was that they not go public with my plight, that they allow the U.S. authorities to work without their having to endure the glare of publicity, in the expectation that behind-the-scenes diplomacy would more readily result in my release than public pressure would. Were there to be publicity, this argument contended, the Russians would hunker down and refuse to do the right thing. Getting me released might take time, Cheri was told, but quiet negotiations were the best way to ensure that I would not be indefinitely held in a Russian prison.

There was a catch in this plan: it depended on the good offices of the State Department. Like me, Cheri was a very loyal American with a bedrock belief in the United States and in American institutions. We are the sort of people who assume that when an American citizen is in trouble, whether or not the trouble is of his or her own making, our government will do its utmost to assist that citizen. Cheri wanted to believe that quiet diplomacy would work and, together with Keith, Brad, the other members of my extended family, and my former colleagues in the NIP community, tried to push the diplomacy process along by bringing pressure to bear on the upper levels of the American government. They all complied with the perceived need to do things quietly. For instance, Admiral Brooks sent an e-mail from the executive board of NIP to the members, advising them that to publicly say the organization was supporting me would only provide ammunition to the Russians for the allegation that I was a spy; instead, Brooks urged NIP members to work as individuals toward my release by writing and calling their congressional representatives.

But quiet diplomacy was nonsense, others in the inner circle advised Cheri. For one thing, they argued — as Cheri remembered me often saying — Russians do not respond properly to courtesy and pleas: you can begin a real conversation with a Russian only after hitting him in the head and making him realize that you have some power and are not averse to using it. Reliance on the State Department and working within authorized channels of communication, they asserted, would get her — and me — nowhere. This group advocated publicity, and lots of it, reasoning that only the pressure provided by the glare of the media would properly galvanize the State Department and the White House.

Both strategies had to be considered, and were, by Cheri, Keith, and the inner circle of advisers. They decided that at this time they really had no choice other than to embrace the quiet, behind-the-scenes approach.

When lawyers Gofshtien and Barannikov showed up at Lefortovo two days after Blubber-Butt had begun to interrogate me, I had no idea who they were, but they told me they had been hired by Keith and Cheri and that was enough to make me happy. Gofshtien was an older, courtly man who knew his way around the system, and the young hotshot Barannikov's English was quite good. Their presence was a great relief, the first and very positive piece of evidence that my family and friends were mobilizing to help. Their entrance on the scene dismayed Blubber-Butt — which also pleased me — and allowed me to get rid of the FSB lawyer who had done nothing on my behalf and who had colluded with Little Feliks and Blubber-Butt in writing protocols that were detrimental to me. "I'm sorry, sir," I told the FSB lawyer, "but I don't know if there is enough money to also have you on my team; if there is, of course I'll be in touch." I had absolutely no intention of ever permitting him near me again, but didn't want to say so, lest he find some way to make reprisals against me.

The chief of the interrogators, Major of Justice Dmitri Shelkov, made my new lawyers sign a paper acknowledging that the proceedings might concern secret/classified material and swearing not to divulge the contents of what was discussed in them to anyone outside the interrogation complex. To have them sign such a document might sound reasonable in terms of defending state security, but it had the intended side effect of becoming a gag order preventing the lawyers from speaking about my case with the news media or to my family and friends, except in the most general terms.

Although passive by American standards, Gofshtien and Barannikov were of moderate assistance in enabling me to insert more clarifying and exculpatory comments into the protocols that Blubber-Butt composed — as my first lawyer had refused to do. Also, because of Barannikov's facility in English, he was able to critique Alyosha's interpreted versions of the questions, my answers, and the language of the protocols, which served as another brake on the interrogator's continual attempts to incriminate me.

Perhaps in reaction to the presence of Gofshtien and Barannikov, on April 14 the prosecutor filed formal charges of espionage against me. The charges themselves did not give us much information because they were nonspecific violations of the espionage statute. My lawyers and I still did not have a handle on precisely what the Russians were accusing me of stealing, or to whom I was supposed to have been passing the information or booty. But the existence of the formal charge meant that there could be no further doubt about the seriousness of the situation: I was the first American to be jailed and charged with espionage since 1960, when U-2 pilot Francis Gary Powers had been shot down during an overflight of the Soviet Union, captured, and subjected to a humiliating public show trial. Powers had been convicted and was incarcerated in Russia for two years before being released in an exchange of spies for the notorious Rudolf Abel. I well remembered that Powers, rather than being celebrated as a hero when he returned home to the U.S., was treated as a pariah. A similar course of events now seemed in store for me, and the prospect was frightening. The Russians took delight in reminding me that violations of the espionage statute carried a ten- to twenty-year sentence if convicted. When they mentioned this, they would snicker like villains in a bad movie.

I wasn't laughing. To be convicted of espionage was no joke, especially since the U.S. government knew very well that I was not a spy, which made it unlikely that I would ever be exchanged for a real Russian spy held in the U.S. I might never get out of prison in Russia!

Blubber-Butt's questioning took off from where Little Feliks's had ended, going over every detail of my time in Russia. The FSB, he said, had been watching me for years and knew all about my "espionage," which he implied had been taking place for several years before my arrest. The assertion that they had surveilled me for years could have been true or it could have been a conceit designed to push me into damaging admissions. I was unable to decide. But Blubber-Butt's questioning did reveal his knowledge of some of my dealings in Russia that preceded my March 2000 visit, and his claim that they had been watching reminded me forcibly that the FSB had indeed shaken its rattles at me before it struck.

A possible warning signal had been sounded on March 17, the day I arrived in Moscow. My plane landed around noon; I went to the Sayani hotel and checked in, and then Bolshov and I proceeded downtown to the offices of Russian Technologies.

As the reader will recall, I was under several contracts for this trip, all dealing with technology conversion. One was from my own company, TechSource Marine Group, two were from ARL at Penn State, a fourth was from ONR, and a fifth was from a private company (whose identity my contract with them still prevents me from naming). That afternoon I was representing TechSource, which had a signed agreement with Russian Technologies for various development projects undertaken by Region, including the authorized transfer of technologies from the Shkval torpedo.

The Shkval torpedo had been touted as able to travel several hundred miles an hour underwater, five times faster than any other torpedo — even faster than a bullet. To achieve such a speed, it made use of a physical principle called supercavitation.

Isaac Newton referred to the basic idea of cavitation in *Philosophiae Naturalis Principia Mathematica,* published in 1687, so the notion was not new. As any swimmer knows, in water you can't move your arms, legs, or torso as quickly as you can move them in air. Water has a pronounced drag, and objects passing through it encounter resistance. Supercavitation can reduce drag by creating a bubble of air or gas in which an object — in this case, a torpedo — can ride, and thus travel much faster than it otherwise could in water. The Russians claimed there were 170 different technologies associated with a Shkval. Our interests centered on only a few of these. The rocket that moves the torpedo, and that also creates the air bubble in which the torpedo rides through the water, is powered by a hydroreactive generator, or HRG. The HRG uses a fuel made from powdered metal, which reacts with seawater let into the body of the torpedo to form a gas; a portion of that gas is funneled back up to the nose of the torpedo and used to create the supercavitation bubble, while the majority of the gas goes out the back end of the torpedo, rocketing the missile through the water.

We wanted to adapt the supercavitation principle, and the HRG propulsion system, for use on a surface vessel — in particular, on high-speed ships being designed and built by the Pequot River Shipworks in Connecticut. Of course, if we could get it to work on these ferries, we could get it to work elsewhere; the market might be huge. A report on an aspect of adapting the cavitation-producing technology to surface-vehicle uses had been given by L. I. Maltzev of the Novosibirsk Institute of Thermophysics at an ONR seminar held at Newport, Rhode Island, in October 1999. Maltzev had arranged for my presence at that seminar and later engaged me as his business partner. He was one of the people I visited in Russia in March of 2000, in order to put him under contract to TechSource for work to be done for ARL and others.

In the TechSource arrangement with Russian Technologies, the other primary participant was Region, a scientific production or manufacturing organization that was also a subsidiary of the Ministry of Defense. I had two main contacts at RT, the head man and an associate named Yuri, with whom I preferred doing business because he was my almost exact counterpart — a former Russian naval intelligence officer — spoke English, and was very good at cutting through the fog that surrounds most business transactions in Russia. Yuri was fine in the March 17 meeting, but the two people who attended from Region were a problem. One of them I knew well: he was the author of an article about high-speed underwater-vehicle technologies that interested us, but he was also something of a mad-scientist type, always going off on tangents and frequently disruptive in meetings. However, since he was the acknowledged expert on the technology, we had to pay attention. The other man from Region had been so quiet in previous meetings that I suspected he might be a security officer. Toward the end of the meeting, the mad scientist asked me to do him a "favor" — obtain a copy of an unclassified publication about hydroballistics published by the U.S. Navy in 1975. It concerned, of all things, the way a missile launched from the air but intended to hit an underwater target reacted when it entered the water environment.

That seemed odd, but I promised to get a copy for him if the 486-page publication was available and authorized for release. Then he brought up a second matter, which disturbed me much more. A few years earlier, before I started TechSource Marine and when I was just an employee of ARL, Region had told us that in order to deal with them (on an entirely different project having nothing to do with the torpedo), we would need permission from their Ministry of Defense. Accordingly, in January of 1997, I had drafted a letter from ARL to the head of the Russian Federation Ministry of Defense unit called Rosvoorouzhenie, the official "state corporation for export and import of armament and military equipment." The letter asked for permission to work with Region on cooperative research concerning conversion to civilian use of certain technologies earlier developed for the military. Ray Hettche signed and sent the letter. By the time I resigned from ARL in June of 1997, there had been no answer to it. A month later, a response arrived in Hettche's office, and it said that Rosvoorouzhenie didn't much care about research, it was concerned with arms sales, and suggested we work with some other subsidiary. So, on behalf of ARL, and later for my own company, in the interim between mid-1997 and 2000, my partners and I had pursued Region and Russian Technologies for several projects and had succeeded in signing agreements with those entities. Now, in this March 17 meeting, the mad scientist from Region charged that our letter to Rosvoorouzhenie had "screwed everything up."

"What letter was that?"

"Ray Hettche's letter."

As the conversation went on, and he asserted that things were now going to be problematic because of this letter, I concluded that either the mad scientist did not know that Rosvoorouzhenie had answered the January 1997 letter, or Ray Hettche had since sent a second letter that I hadn't known about. At the end of the meeting with the Region and Russian Technologies representatives, I raced back to my hotel, concerned and confused. It was then I noticed that the phone head-piece in my room had been changed, an obvious sign that my telephone was being tapped.

But despite the probable tap on the phone, I was so steamed up that I had to find out if a second letter had been sent from Hettche to Rosvoorouzhenie or Region. Was ARL trying to cut a deal behind my back? Was the mad scientist's mention of the letter a warning sign of things going wrong? Why had Yuri's boss been absent from the meeting — was he trying to distance himself from me? Was the odd request for a 1975 Navy publication part of a plot to get me to do something I didn't want to do?

At ARL, a secretary took my call, and after two hours Ray's deputy Ed Liszka called me back from San Diego, where he was attending a conference; Liszka reported the gist of his own conversation with Hettche: that no second letter had been sent.

That relieved only some of my anxiety. Dan Kiely was to come to Russia to provide technical expertise on the project I was doing for ARL, and I reminded Liszka of the need for Kiely — for all of us — to remember that this was Russia, a dangerous place. Kiely, I warned, would have to be damned careful about what he discussed while he was here, and circumspect in regard to what papers he carried into the country.

"If there's any difficulty, maybe we shouldn't send Dan," Liszka said.

"Now that I know there was no second letter, the difficulty seems to have been simply a misunderstanding, and I believe everything is okay here. However, if you don't want Dan to come, that's your call," I told Liszka. I further explained that I did not absolutely require Kiely's expertise right then and there, that I could bring home the new documents and Dan could pose questions about them from State College rather than from inside Russia. For almost an hour on the phone, we went around and around on this. I instructed Liszka that Kiely must not bring sensitive papers with him and must not have sensitive files in his computer. By "sensitive," I meant such documents as the five ARL–Bauman University technical reports that had been signed in previous years, preliminary proposals for future arrangements that were still in the process of being negotiated, and papers dealing with other ARL projects (such as direct contracts from the Navy) that had nothing to do with the trip to Russia but which Dan might take along to work on during his leisure hours — as I'd seen him do many times. I stressed that we could not be blind to the fact that we were working in a delicate area, the conversion for civilian purposes of a technology heretofore used solely for defense — the propulsion system of a torpedo. Therefore, we must not even inadvertently convey to the Russians the impression that our interest was connected to military applications. Region had told us they were working on a new version of the Shkval torpedo that had better guidance and control systems, and that this new version was off-limits. We were interested in the propulsion system of the old torpedo, so we had told them that the new torpedo's being off-limits was not a difficulty for us — but I could easily imagine the KGB misconstruing our real interest and I didn't want to have documents around that might feed their paranoia.

"What if you get caught?" Liszka worried.

"Caught doing what?" I countered. "Everything we're doing is legal and aboveboard. We just need to be careful."

"Maybe Dan shouldn't go."

"Again, that's your call."

Two hours later, Kiely himself phoned me from John F. Kennedy Airport in New York, just prior to boarding the plane for his trip to Russia. I reiterated the warning signals and went through the whole litany of what sorts of materials ought to be left home. He, too, asked whether he ought to come, and I told him it was his choice, but that he must exercise caution.

Dan said that Liszka and Hettche had advised him not to go, but he had decided in spite of the warnings to get on the plane and would of course be cautious. It wasn't his first trip to Russia, and he believed he knew how to take care of himself there.

I worried nevertheless, because Kiely was the archetypal absent-minded professor, brilliant in his own field but lacking common sense: I had seen him forget to tie his shoelaces or to zip up his pants after going to the bathroom. He'd eat powdered doughnuts and then wipe his hands on his shirt, oblivious to the telltale stain he was making. He knew A to Z about propulsion systems and thermodynamics, but in conversations he couldn't stick to the subject at hand and would sometimes boast about other projects he had done for the Navy. Kiely's sloppiness, combined with the delicacy of our field of interest, would prove dangerous to us both in Russia, and disastrous for me.

In my early sessions with Blubber-Butt, the interrogator had confronted me with information about my prior dealings in Russia, information whose origin I did not then know but which I began to suspect could be papers that, despite my warnings, Kiely had brought with him. Later on, I would discover this was precisely so: the indictment would focus on five reports that I had "improperly" received from Babkin at Bauman, along with an "incriminating" white paper that I had written, and several other memos with my name on them. I had had none of those documents with me! They were actually all innocuous and legal, but they were sensitive, and I had left them home on purpose and had beseeched Kiely to do the same. Unbeknownst to me, Kiely had brought along copies of the five Babkin papers. As for the white paper, an unfinished proposal still being negotiated between my company and Russian Technologies, Dan thought he had erased it from his computer by putting it in the "trash." However, as anyone who is slightly sophisticated in the use of computers knows, a file thrown into the trash does not entirely disappear but continues to exist on the hard drive and can be extracted and read by a determined investigator.

As I would later learn, that wasn't all that Dan Kiely had provided, wittingly or unwittingly, to the FSB that would be used to damage me.

It did occur to me during my interrogation that my March 17 phone conversations with Liszka and Kiely, in which I had pleaded with them not to have Dan carry the sensitive papers, could well have been construed by the FSB — listening through the phone tap — as conspiratorial, a red flag waved in the face of a bull.

The second series of interrogations was conducted by Blubber-Butt on a less demanding schedule. Generally, there were no sessions on Mondays, and none or only a short one on Fridays. The weekends were also free of interrogations. The sessions did not start until ten in the morning, and around noon were interrupted for an hour or two during which I was taken back to my cell for lunch. Afternoon interrogations would end before dinner. If my lawyers were unable to attend on a particular day, no questioning would take place. Delays were usually welcomed by the interrogators, who did not like to work very hard, but at one point the FSB tried to sanction my lawyers when they did not make it to Lefortovo for a session they had previously agreed to attend. The session was not held, and Blubber-Butt asked me to sign a protocol that said it had not been held because of my lawyers' absence. I wouldn't do that, and asked permission to telephone the lawyers. After some scurrying about and Blubber-Butt conferences with Shelkov, permission was denied. I still refused to sign. Blubber-Butt fumed and turned red, then had me taken back to my cell.

What with all the downtime, I spent a lot of hours in cell number 82. My cell mate Sasha and I became wary friends. He was in his late twenties, unmarried, but had a girlfriend and a daughter. To him, detention seemed an unpleasant but not unexpected interlude between engagements; after all, more than one quarter of all Russian men spent some time in detention. Sasha stood accused of counterfeiting. He loved cars and was forever perusing German and American car magazines. He taught me some Russian, and I taught him some English, and at times we would be jointly defiant of the authorities. After he taught me the Russian term for crazy house, *durdom*, I lettered the phrase in Cyrillic on a piece of paper and we pasted it up in our cell; it stayed there for the better part of a month, until our next cell inspection, when the sign was taken away as evidence of our incorrigible antiestablishment behavior. Sasha also helped me figure out how to use the library: you requested books from their list, and two would be delivered; two weeks later you could exchange those for two others. Amazingly, there was a decent selection of English-language volumes to choose from, though the collection included numerous old Communist propaganda sheets. The first novel I borrowed from the prison library was Larry McMurtry's *Streets of Laredo,* which I thoroughly enjoyed; I also became ensnared in a phrase McMurtry had a character employ: "a nickel's worth of dog-shit." To pass the time in the cell I wrote a thirty-two-page screed on this subject. When Brad Johnson delivered to me a copy of the Bible, I read it repeatedly and thoroughly, underlining passages that were particularly relevant to my situation. I started more regular prayer, too. Of particular solace to me was Psalm 144, verses 7 and 8, which I read every day:

> *Stretch out Your hand from above;*
> *Rescue me and deliver me out of great waters,*
> *From the hand of the foreigners,*
> *Whose mouth speaks lying words,*
> *And whose right hand is a right hand of falsehood.*

Mindful of the possibility that Sasha was an FSB plant, I felt he could be useful to me, and he was. In addition to Russian lessons, Sasha offered me some of the food he had received, a welcome change from our awful daily diet. Every morning we would be given kasha porridge and tea; at midday it was the big meal of cabbage soup, steamed fish, and mashed potatoes; and in the evening, more kasha porridge. Black bread came every day; it was almost inedible. The white bread delivered three days a week was better. One hour each day, we prisoners were taken up, a cell at a time, to a rooftop exercise area open to the sky and surrounded by high walls; it was not much bigger than a normal cell, but it was empty of fixtures, and we could walk unrestricted in it for an hour; I usually walked the equivalent of three miles. After sweating during our exercise, we took turns sponging off in the sink of the cell, until Thursdays, when we were taken out of the cell to the steam-bath rooms and allowed a once-a-week cleansing. I tried to maintain muscle tone by doing sit-ups and push-ups in the cell.

Sasha told me stories of his participation in the *Bratva*, the brotherhood, or the Russian Mafia — wild tales of murders, prostitution, and protection rackets; proud of his affiliation, he viewed his arrest as an occupational hazard that he would soon overcome. Sasha's Mafia tales could have been pump priming, told in expectation that I would reciprocate and regale him with my own illicit behavior. He might not get me to admit to espionage, but might push me into boasting of having a Svetlana on the side, or of conducting other nefarious activities that I didn't want my wife or business associates to know about — ammunition the FSB could use to blackmail me into a confession or a guilty plea. I ate Sasha's food but didn't bite that particular apple.

On April 21, a new man was put into our cell. Fat and gross, a chain smoker and incessant talker, Nick was also an incurable snorer. The noise that he made was so loud that it was impossible to sleep when he was in dreamland. Sasha and I had to try to adjust our schedules, to sleep when Nick was awake, and Nick cooperated by dozing when we were awake or out of the cell. I was certain that Nick was an FSB plant because he would be taken out of the cell for interrogation and not returned for quite a long time, more time than was being spent on my interrogations. He chattered continually in Russian at Sasha, even when Sasha pointedly interposed his magazine between himself and Nick's face; sometimes Nick would move his position to try to force Sasha to pay attention to his monologues. I wondered if Nick had been placed in the cell as a contrast to Sasha, to annoy us both and drive me into closer alliance with my first roommate. I had no proof of these suspicions, but Nick was soon removed from the cell, and we were told that he had been released from prison.

One of my worst days in Lefortovo was May 2, Cheri's and my thirtieth wedding anniversary. In recent years, I had been spending so much time going back and forth to Russia, and becoming consumed by my new career as a businessman, that I knew Cheri felt that I had not devoted enough time to being with her. The thirtieth anniversary was supposed to have been an occasion to make up for lost time. We had determined to go off by ourselves to Cancún or some other exotic vacation spot but had put off a decision on the exact destination, reasoning that we would finalize our plans after my return to the U.S. Now we couldn't even communicate. I knew she was thinking of me as I was thinking of her. I felt somewhat sorry for myself but felt a lot worse for her: I was causing her pain, something that I had never wanted to do. And I couldn't stop her pain.

Back home, something important was happening with Cheri, which I knew nothing about: her application for a visa to visit me had first been granted by Russia and then summarily "annulled" as she had been on the verge of packing her bags. This was highly unusual, especially since she was traveling to Russia under the auspices of Congressman Peterson — such visas are almost never denied. But the government of the Russian Federation did not want Cheri Pope in Moscow just then, and the U.S. government seemed unwilling or unable to help her get there.

Although our State Department people later insisted that this was just a bureaucratic snafu, the FSB had good reason to keep Cheri out of Russia at that moment: without her there, they could monopolize the news about Edmond Pope. The Moscow newspapers, television, and radio stations were pushing an FSB version of my case under such headlines as "Russia Pressing Issues — American Spy Tried to Get Secrets of Our Super-Missile." Although false, the media account of the case was helpful to me, as it made specific what had previously been a general accusation: the Russians were charging me with stealing the "secrets" of the Shkval torpedo. The media account also said that the FSB had a videotape of me receiving drawings from my Russian contact and giving him money, part of a payoff of $30,000 that I was supposed to have made.

The Shkval had been for sale to foreign buyers for years, so the charge of stealing its secrets was nonsense. I wondered what the FSB video could possibly contain. As for the $30,000, that had some relationship to the amounts that ARL had paid to Bauman University in

1997, but none to any money I had disbursed on this most recent trip. Oddly, the media account of my espionage included the tidbit that I had obtained receipts for the money I had turned over; to anyone familiar with factual or fictional spy operations, that should have raised eyebrows, for what spy ever asked for a receipt for an illicit transaction?*

Predictably, most of the Russian press had made no attempt to balance its reportage by obtaining my side of the story, either from me or from my lawyers, as would have been routinely done in the West. The media refusal to investigate both sides, the constraints placed on my lawyers regarding discussion of the specifics of the case, and the official government refusal to let Cheri into the country (or near a Russian television news camera and microphone), left the FSB in full control of what the Russian public was permitted to think about me.

Blubber-Butt's interrogation reached a potentially treacherous area: the close examination of my years in the Naval Intelligence Service. I had much to conceal here, not the least of it the many classified programs for which I had been cleared. In Lefortovo, I did not even know the precise number of those classified projects and thought it was around 50; I later learned it was 126. Although some of the classified information in my head was out of date, other bits and pieces of information, should they come into Russian hands, could damage U.S. security and interests. There were even aspects of the old Whiskey-on-the-Rocks incident that were still classified. Beyond

*On April 6, when the FSB released its initial statement that an "American spy" had been arrested, and had receipts for cash payments to Russians in his possession, retired KGB officer Konstantin Preobrazhensky told a Moscow newspaper that "it 'looked strange' that the American was carrying receipts for cash payments, something professional spies try not to do. This, he said, may indicate that the American was doing research on his own rather than spying." *Moscow Times*, "FSB Arrests U.S. Citizen Suspected of Spying," April 6, 2000.

that, I was mindful of the Russian proverb "There is no such thing as a former spy," and expected Blubber-Butt's attitude to mirror the thought behind that aphorism.

My defense against his questions about my years in Naval Intelligence was to treat them as unimportant, to play at being nonchalant and deflect his inquiries with partial answers. "Oh, the Gulf War? Yes, I ran the briefing office. We received field reports, made digests, and briefed the admirals and other big shots on what was happening." What I said was true but far short of the mark. Any competent interrogator with the faintest suspicion that I might be lying would have bored in, asked many other questions, forced me to be specific, tripped me up with the apparent inconsistencies in my answers — and thereby pushed me to reveal secrets of classified weapons systems and intelligence capabilities.

What they should have done was bring in their own experts, Russians who knew some things about American secrets, and have them prod me for details. What about that unmanned high-glider vehicle? The high-speed data link to the ground? The development of this or that ordnance delivery system? Tough questioning by experts on these subjects, extracting some hitherto-concealed information, could have done me, and the United States, a lot of damage. But no experts were brought in for the interrogations. None. Had they been, I would not have cooperated, because I'd decided at the outset of my detention that the Russians would get nothing of consequence out of me willingly.

The combination of the FSB's readiness to accept my bland explanations for my Naval Intelligence days, of Kiely's being released but me held, of Cheri's being denied a visa, and of the spin that the Moscow newspaper and television reports were putting on my case made it clear to me in prison, and also to my family and friends at home, that my arrest and charging for espionage were politically motivated. Most people suspected by the Russians of espionage in recent years had been detained for a day or two and then deported, rather than arrested, held for long periods of time, and sent to trial.

I was not being treated in a perfunctory manner, which made it clear to me that the timing and emphasis of my arrest were tied to something. That something was none other than the very recent election of the new president of the Russian Federation, Vladimir Putin.

CHAPTER 4

The Reasons Behind My Arrest

A month after I began with Blubber-Butt, Sasha and I were transferred to a double cell that accommodated six men. During the next two weeks, many detainees were shifted in and out: older and younger, sophisticated and simple Russians; men from neighboring republics who stood accused of terrorism; three black men from Africa arrested while trying to leave Russia on friends' passports. Though disparate in origin and temperament, all of us came to share the camaraderie of the oppressed, to distribute among us the food packages received from outside, and to try to accommodate our idiosyncrasies so we could exist as pleasantly as possible in the small space. One matter we agreed upon wholeheartedly was distaste for the just-elected Vladimir V. Putin. If the reactionaries in the KGB, the military, and the Russian Mafia had searched far and wide to find a figurehead behind whom to take back power from the pro-democracy forces, they could not have dreamed up a better candidate.

In August of 1999, the world's news organizations reported that ailing Russian Federation president Boris Yeltsin had once again replaced his prime minister. Yeltsin had made several such switches in recent years, so not much notice was given to this event. The public explanation provided for the change was that Yeltsin had decided that to better fight off political rivals like Moscow mayor Yuri Luzkhov, he needed to replace Sergei Stepashin, appointed as prime minister only in May, with a new man and probable successor. The man chosen was the forty-six-year-old Putin, who since July of 1998 had been serving as the head of the FSB.

Questions immediately arose as to whether Putin had been forced on Yeltsin by the powerful cabal known as "the Family," since Putin was in such fundamental disagreement with Yeltsin over many matters, chief among them the worth of the state security apparatus. A centerpiece of Yeltsin's rule had been the deliberate dismantling of the KGB. Putin had attended KGB law school, worked for the KGB for sixteen years, and still frequently made statements lauding the integrity of state security agents while calling such men as Oleg Kalugin, a former KGB officer who had joined the radical pro-democracy forces, a traitor and an "absolute loafer."

At age fifteen, in 1968, Putin had watched the movie *The Sword and the Shield,* about a KGB double agent in Nazi Germany, and was emboldened to march right up to the KGB's Leningrad headquarters and volunteer to join. An older man there advised him to first go to college and law school, and then apply. Dutifully, Putin attended college and then the KGB's law school, and also became a master at judo, before joining the organization in 1973.* Of his sixteen years with the KGB, he spent the last half dozen in East Germany, before leaving the organization in 1989 to go into politics. As Leningrad returned to its original name of St. Petersburg, Putin rose in the hierarchy of the mayor's office. A few scandals were associated with his tenure, but they were not serious enough to hamper his transfer in the mid-1990s to the Russian Federation bureaucracy in Moscow. At the Kremlin he worked as liaison to the regions and in other responsible posts before rising to head of the FSB. During his tenure at the FSB,

*A documentary aired on Russian television while I was in Lefortovo claimed that Putin had once worked at the Krylov Shipbuilding Institute; I had visited there many times, made some friends among the employees, and had an ongoing research contract with the institute.

several dubious cases of espionage were brought against people involved in monitoring environmental pollution in Russia, and Putin commented in an interview that environmental agencies ought to be viewed as tools of foreign-intelligence services.

When Putin was appointed prime minister in August 1999, one of his main tasks was figuring out how to deal with the breakaway republic of Chechnya, whose forces had humiliated the Russian military several years earlier. "The first Chechen War was lost because of the moral condition of society," he said, and against the recommendations of four recent prime ministers, Putin agreed with the generals on the need for an all-out surprise attack on the Chechen capital of Grozny. The trigger for that attack was a series of bombings that devastated several apartment buildings in Moscow, killing more than 300 people. The bombings were blamed on Chechen rebels, although some analysts believe them to have been the work of Russian military and state security apparatchiks seeking to build greater public support for widening the war. The culprits have never been identified. Several of my cell mates felt certain that Russian security forces had indeed been responsible for these bombings.*

Grozny, a city of 400,000 people, was reduced to rubble. But almost immediately after the attack on Grozny began, Putin's standing shot up in the polls, and he was soon considered a fit successor to Yeltsin, who was scheduled to step down from the presidency in midyear 2000.

As Russia headed into the parliamentary elections, Putin had three strong rivals for the post of president: Gennadi Zyuganov, leader of the Communist Party; the centrist former prime minister Yevgeny Primakov; and a right-wing former prime minister. Primakov was considered the most potent, so during the parliamentary election campaign, he and his Fatherland–All Russia Party were subjected to repeated punishing attacks by television stations controlled by a

*Four cell mates were under investigation for participation in "terrorist plots." They believed their real crime was that they looked suspicious.

principal oligarch in the Family, which had long supported Yeltsin and was now helping protégé Putin. Primakov's party lost the parliamentary election to a new Unity Party that had made an alliance with the Communists in order to amass power. Shortly, Unity announced it would back Putin for the presidency. It was this show of political power, many observers believe, that catalyzed Yeltsin's unexpected retirement from the presidency on December 31, 1999, and the announcement that Putin would become the Acting President as well as prime minister. Little noticed by the West, upon his retirement Yeltsin was simultaneously pardoned by the new Acting President for all past crimes, whether or not they had yet been discovered. As required by the Russian Federation constitution, elections for the presidency had to be scheduled within three months of the previous president's resignation; the precise date set was Saturday, March 26, 2000.

In January, Putin and the Unity Party took control of the Duma's leadership positions, shutting out Primakov, who had sought the speakership. Primakov soon dropped out of the race for the presidency, telling reporters, "I sense how far our society is from being a civil society, and from a true democracy." American financier and philanthropist George Soros, who had contributed more than $100 million to rebuild Russia from the Communist-era ashes and was a close student of the Russian Federation, opined that "the state built by Putin is unlikely to be based on the principles of an open society." Few in Russia were paying attention to Primakov, and Soros's operation was considered by the FSB and the public it influenced as no more than a cover for Western espionage — the same charge leveled in the past against many aid programs.

But Soros and Primakov were right on target. During the months that Putin was Acting President, he signed Decree Number 24, which beefed up security internally and mandated a strengthening of Russia's international position, for instance by the resupply of naval bases in Vietnam and Syria that had been all but dormant because they lacked ammunition and other essential supplies. A companion "Reform of the Presidential Administration" aimed to once again subordinate all political life in Russia to central control and recommended the use of force where necessary to exercise this control. Putin and his backers also decided to resume arms sales to China, Iran, and other former client states. The Iran sales had been specifically contravened by the Gore-Chernomyrdin protocol of 1995, in which the Yeltsin government had agreed to gradually stop selling weaponry to Iran and to halt the sales entirely by the end of 1999. According to a Russian military newspaper, *Krasnaya Zvezda,* "The U. S. was supposed to compensate for our losses [in arms sales to Iran and China] by helping to promote Russian weaponry in other markets," but it had not done that. For instance, the U.S. had promised to help the city of Kharkiv, in Ukraine, if it agreed not to help Iran build a nuclear reactor. Kharkiv had agreed, and endured serious economic consequences, but Iran went ahead with the nuclear plant with help from other sources, and the promised U.S. aid for Kharkiv never materialized. The Putin government was said to have no choice but to resume arms sales soon. Behind the scenes, it had already completed an agreement with India to provide advanced weapons for $2 billion.

The official presidential campaign season opened on February 26. Putin promised to bring a market economy and the rule of law to a federation whose main problems he identified as "lack of will" and "lack of firmness," asserting that "the stronger the state, the stronger the individual." A full-page advertisement of his tenets, in *Izvestia,* was undermined that day by a videotape aired on many European stations that showed Russian troops conducting a mass burial of Chechen troops; some of the victims were mutilated and others were bound by restraints, as though they had been shot while held in detention rather than in battle. This revelation of Russian military excesses by foreign journalists was dismissed by Putin's government as unwarranted interference in internal Russian affairs — a response reminiscent of explanations given during the Communist era. The war in Chechnya remained popular with the Russian people, in part because of relentless propaganda from the Berezovsky media empire. The Chechnyan campaign was billed as a test of the Russian Federation's renewed manhood and of its ability to exert its weight and influence in the wider world.

Putin's election campaign was run by Berezovsky's public-relations guru, who tried to distance Putin from the excesses of the Yeltsin era, for instance contending that "there are members of the old oligarchic circles, others from the regional elite and a significant part of the old Yeltsin apparatus who fear losing their posts and old corrupt ways should Putin come to power." Putin himself, in interviews for a book, characterized the 1990s and their democratic reforms as a decade of disaster and humiliation that needed to be rejected in order for a strong Russia to reemerge.

There were other warning signs that a duly elected Putin would be considerably more authoritarian than Yeltsin and would restore to full power the KGB-successor agencies that Yeltsin had decimated and dispersed. Intimidation of press organs that criticized the acting president was coupled with threats to deny license renewals to the lucrative television stations owned by Gusinsky — stations that in response quieted their criticism and promoted Putin's election. When the well-regarded *Kukli* comedy puppet show on NTV depicted Putin as a surgeon with a blowtorch and a hatchet as his instruments, public rebukes followed — as they had not done during the prior six years of the television show's existence. Another tactic used by Putin's backers was direct grants from the Kremlin to news media in the outlying regions, making those outlets more responsive to propaganda from Putin's campaign.

As the month of March began, it became clear that the election would be little more than a coronation. The war was popular, the major media had been coopted, the liberals were fragmented, and the Communists had few media outlets with which to spread their message. Putin declared himself above politics, but, concerned that he might not win 50 percent of the votes on the first ballot and would thus have to enter into a runoff, he made promises to specific sections of the electorate. To the military, he promised bonuses of three times the usual size; to state employees, a 20 percent wage increase; to ordinary Russians, a rollback of the high tax on vodka that had been imposed under Yeltsin. This last promise was reminiscent of one made by Putin's hero Yuri Andropov when that KGB chief became the leader of the U.S.S.R. in 1982 and won the hearts of ordinary Russians by decreeing cheap vodka. On the international front, Putin promised the London Club (commercial creditors) and the Paris Club (governmental creditors) to reschedule Russia's debts, on which Russia had been in default since August 1998. He also rattled a saber, having the Ministry of Defense charge that an American radar station, sixty kilometers away from Russian soil in Norway, was spying on Russian ballistic missile tests. The U.S. response — that the station had been in operation since 1995 and was used to spot man-made debris in space so that it could be redirected to fall on unpopulated areas — was met by official Russian derision.

Putin's public appearances stressed the need for central control over the regions, and he replaced one fourth of the federation's representatives in the regions with people more inclined to assert that central control, many of them his former colleagues in the Leningrad KGB. He also laid plans to undermine regional authority by preparing a decision in which the courts were to be permitted to delay the enforcement of local executive regulations, including gubernatorial decrees.

On March 26, 2000, Putin won election to the presidency by an official 52 percent of the vote, enough to forestall a runoff. Communist Party poll watchers and other critics reported massive fraud in the election, but their protests rang hollow, for the Communists themselves had used the same tactics in the past; also, nothing changed the election results. Putin, now elected, moved quickly to radically alter the direction and control of the federation, taking steps reminiscent of Hitler's consolidation of power in 1933 after being elected chancellor of Germany.

He appointed two KGB Ivanovs to positions of significant responsibility: Sergei Ivanov, a career state security officer, became head of Putin's security council and the unofficial vice president, and Viktor Ivanov, another long-serving KGB officer, was made head of the Kremlin's personnel department. Nikolai Patrusev, who had succeeded Putin as head of the FSB, would shortly be given control over the prosecution of the war in Chechnya, overseeing the entire military hierarchy.

These and other KGB/FSB stalwarts in the Putin inner circle had been tarred with the recent failure of a case of a Russian suspected of espionage. It was against Aleksandr Nikitin, a retired Soviet Navy captain who helped a Norwegian group document the dumping of nuclear waste in the Arctic seas. International public pressure helped a St. Petersburg court find Nikitin innocent, and, in a blow to the prestige of the FSB, to rule that the FSB had acted illegally in gathering evidence. Toward the end of March 2000, the FSB's appeal of that verdict was about to be heard in the Supreme Court when the hearing was suddenly postponed indefinitely, probably because the Supreme Court was about to sustain the lower court's finding of innocence. After such a humiliation, and the failure to halt terrorist activities, Putin's inner circle and the FSB certainly needed to find a good case against a new spy.

Two days after Putin's electoral victory, my arrest was authorized for the FSB by the general prosecutor's office. Did Putin personally order my arrest? Many aspects of the arrest argue that he was involved. The first is the timing.

In the wake of his electoral victory, Putin aggressively abandoned whatever restraint he had professed while on the campaign trail. Not only did he encourage more active harassment of media tycoon Vladimir Gusinsky, but he even went after the Berezovsky media empire. The promised rescheduling of foreign debt fell by the wayside;

dialogues with human-rights groups within Russia, which had been held frequently by the Yeltsin government, were shut down; press freedoms were curtailed; the eighty-nine regions were divided into seven super-regions whose headquarter cities were the same as those of the seven military districts of the Russian Federation. The Putin administration also stepped up attempts to finesse the Gore-Chernomyrdin agreement barring arms sales to Iran so that they could once again sell munitions to the ayatollahs, and to raise the hard-currency amount of arms sales to China — including sales of the Shkval torpedo.

If Putin and his backers wanted to send a signal to the Russian peoples that Russia would no longer be pushed around by the United States and would resist being relegated to a second-order power, there was no better signal than to arrest and charge with espionage an American businessman and former Naval Intelligence officer who had been working to purchase elements of technologies that were derived from and associated with advanced Russian weaponry.

My arrest also distracted Putin's domestic audience from the war in Chechnya, which was not going well, and focused its attention on the need to be as vigilant against spies from the West as it had been in Soviet times. Two prominent former KGB people understood this message right away. KGB general Oleg Kalugin, who had led Yeltsin's dismantling of the KGB and whom Putin had called a traitor, finally decided to settle permanently in the West after Putin's election. He wrote an open letter to British and Russian newspapers stating that he would not return to the Russia of Putin, "criminal and corrupt, with its pocket justice and court system." The second person to speak out was the author of a book on the FSB and a former lieutenant colonel in the KGB. He wrote in the *Moscow Times* that "the hounding of Kalugin" meant nothing less than

> the resurrection of the KGB. This is the security services' vengeance for a decade of humiliation by democrats. . . . A newly resurgent KGB will use the old KGB methods, creating a cult of secrecy, "imposing order" in dealing with foreigners, trips

> abroad. . . . Dissidents will appear once again. That's just what the security services need, someone to work on, to compromise, to arrest.*

Further, my arrest signaled the Russians working with advanced military-developed technologies to stop being so eager to make deals with American businesses, lest they too fall afoul of the FSB. The American "partners" of those Russian scientists were similarly warned to back away from further attempts to buy (cheaply, in the Russian government's estimation) what had taken Soviet science many years and countless hours of manpower to produce. Last, but surely not least, my arrest sent a signal to President Bill Clinton — whom Putin detested for what he saw as Clinton's lack of moral fiber — that this new Russian president would not be a pushover for the American government, as the hard-liners in the Kremlin believed Boris Yeltsin had been.

In January 2001, in an article by reporter John Mintz of the *Washington Post*, unnamed but knowledgeable sources were quoted as claiming that my arrest had not been caused by anything I had done, but had occurred because "Pope fell afoul of an intelligence operation in which he was not involved: an effort by the Canadian government to buy a handful of Russia's advanced Shkval (or Squall) torpedoes from a defense plant in the former Soviet republic of Kyrgyzstan."† Mintz suggested that the Canadian navy, which had limited possible uses for such a torpedo, was probably buying it on behalf of the United States of America. The *Toronto Star* later wrested an admission from a high-placed official in the Canadian military that they had indeed tried to buy a Shkval; the official also acknowledged that

*Konstantin Preobrazhensky, "Resurgent KGB," *Moscow Times*, April 18, 2000.

†John Mintz, "Unseen Perils in a Russian Squall: How Edmond Pope Fell Victim to Intrigue Over a Torpedo," *Washington Post*, January 3, 2001, p. A1.

Canada was looking into sharing any information obtained from the purchased Shkval with the United States.*

I have since spoken with the sources of Mintz's article, as well as others who were aware of certain aspects of the situation. Some turned out to be individuals with whom I had previously crossed paths in mutual pursuit of technologies made in Russia and other overseas business opportunities. They were not arms dealers in the sense of being manufacturers like Glock who sell pistols, or middlemen who procure boxcar loads of Kalashnikov rifles for Third World countries; rather, they were dealers/contractors to the various military services of the West, who had been using them for years to purchase examples of Russian weaponry in order that the West's military services could test the weapons' capabilities and then design systems to counteract those weapons. According to their sources in Russia, there was a contract with the Canadian government, in which other Western intelligence agencies participated, to pursue the purchase of Shkvals from a test facility in Kyrgyzstan.

These dealers/contractors were sophisticated people, used to doing delicate business in Russia. Some admitted to having tried to protect themselves from arrest by making "arrangements" with the Russian Mafia and the FSB, but said that in all other respects their dealings in the former Soviet Union had been within the confines of American and international law. The deal one group set up for the Shkval was not their first for an advanced Russian weapon system, they said, and it was complicated, as are all deals that involve military technologies from the former Soviet Union. It encompassed the participation of several republics of the former Soviet Union, the Region Design Bureau, Rosvoorouzhenie, and a test facility in Kyrgyzstan from which the torpedoes were going to be obtained, with — as several sources told me — the red and white test stripes still on but with the nuclear warheads removed. Twelve million dollars were to change hands in the deal.

*Allan Thompson, "Canada Wants to Test Russian Squall Torpedo, Eggleton Confirms," *Toronto Star,* January 5, 2001, p. A1.

While the contractors believed they were being aboveboard in their dealings, they were aware that Rosvoorouzhenie was frequently referred to in Russia as "*Rosvori,*" which literally means "Russian thieves." It was also known as the finance arm of the FSB, and the contractors understood implicitly that the money pie that they delivered was to be cut up in many slices, at least one of which would go to the FSB and another to the Russian Mafia.

Generally, when such dealers/contractors purchase a weapon, they do so without being able to obtain along with the weapon the necessary documentation for it, since that documentation usually stays within the control of the military or the FSB. At first this might seem strange, but it actually makes sense. Reverse engineering of a complex technology is a lot harder than most people realize. If a Russian were to sell a torpedo with the plans, it would be much easier for the buyer to learn all he wanted to. But to buy a torpedo without the plans and documentation was — in this view — like buying a computer without an operating manual, and it carried the risk that the torpedo might prove useless, not understandable enough for countermeasures to be designed. Several of my sources believed that in a roundabout scheme the Navy had cleverly engaged me, a private businessman, to solicit information from precisely the people in Russia who could supply the documentation on the Shkval. Of course, my contract for the HRG work had been with Penn State's Applied Research Laboratory. But the impetus for the project, the dealers insisted, had come from the Navy, the ARL's major source of funding.

This explanation is not a perfect fit, since my work for ARL with Bauman University began in 1997, likely well before the contract with the Canadian Defense Ministry was inked — but still, I had been working in Russia with the people from whom documentation critical to testing a Shkval might be obtained. It must have appeared to the Russians at the time that I had been sent out to buy for the Navy everything associated with the Shkval but the hardware. In pursuing detailed technical data on the Shkval, in other words, I was the perfect spy — one completely unaware that I was "spying."

But the contractors' deal for the Shkval went sour in March 2000. Somebody on the selling side did not pay a "commission" to the FSB, or to the right people in Region or Rosvoorouzhenie. The dealers/contractors lost their money and their contract.

Soon after the deal went bust, according to the sources, either in a case of mistaken identity or simple FSB revenge for not having received its cut of the deal, I was arrested by the FSB and charged with stealing the "blueprints" of the Shkval. Of course, I *was* involved in purchasing something associated with Shkvals, but for a paltry sum, merely tens of thousands of dollars, while the dealers' contract was for $12 million. It would have been easy for a higher-up in the Ministry of Defense, the Mafia, or the FSB to conclude that I was trying to get away with buying a Shkval for pennies on the dollar, and for him to have taken angry action against me to prevent that too-small deal from going through and depriving him of a cut of the larger one.

The people to whom I spoke, who operate regularly in Russia, point to the nine days that elapsed between my arrest and the lodging of a formal charge against me as evidence that the FSB was expecting and hoping that someone in the U.S. government would get the message and realize that a payoff was due to the FSB for the "commission" it had forgone on the test-Shkval deal. Had the lost commission money been paid to Rosvoorouzhenie during that nine days, they allege, I would not have been charged but would have been released.

Within several weeks of my arrest, the Russian Ministry of Defense did completely reorganize its entities involved in arms sales, revoking the charter of Russian Technologies, with which I had been dealing and which was now deemed to be in too-vigorous competition with Rosvoorouzhenie. Later in the year, that entity too was disbanded, and all potential foreign buyers of Russian arms and technology were henceforth told to go through a new organization directly under the Ministry of Defense. In an article printed during my imprisonment, and well after Russian Technologies had been disbanded, a Russian defense analyst told the Cox newspapers that my arrest could have been a by-product of a battle between Russian weapons companies, and "he could have mistakenly landed in the middle of something over his head."*

As an explanation for my arrest, the scenario of a secret deal gone awry is plausible, but my decades as an analyzer of information make me chary of awarding it 100 percent reliability. More likely, it was one among several reasons that the FSB zeroed in on me. This scenario presumes that I was either too naive or too stupid to realize I was heading into trouble on this particular trip. Although I don't agree with that assessment, I will certainly cite my own shortcomings as among the possible reasons for my arrest. Perhaps I should have been more suspicious and cut the trip short, or not gone to Russia just then. But I had a firm belief, based on the permissions we had been given, that we were doing nothing wrong in our dealings with the Russians, and so I discounted the intensity of the danger signals. I would never have made any progress in Russia had I not been willing to tread a difficult path. And, of course, at the time of my March–April 2000 trip to Russia I knew absolutely nothing about any secret operations related to the Shkval. Had I known or even suspected that I might be being used, I would certainly have stayed home in Pennsylvania.

Two months after Putin's election victory, President Bill Clinton was scheduled to arrive in Moscow for a summit with the Russian president. It was to be their first official summit, though they had met twice between the time Putin had been selected as heir apparent by Yeltsin and his election as Russian Federation president. Sitting in

*Margaret Coker, "Alleged U.S. Spy Jailed in Russia Raises Cold War Fears," Cox News Service, June 21, 2000.

Lefortovo, I entertained the hope that as a result of this June summit I might be released. There were extensive rumors about such a release, even articles to that effect in Russian papers. Here is the gist of one, from *Obshchaya Gazetta:*

> Former U.S. spy Edmond Pope is likely to be released before Bill Clinton's visit to Russia, since his activities in Russia have done no significant damage to Russia's defense capacity. The "top secret" weapon he was investigating was actually one of Russia's better-known torpedoes, authorized for export. . . . It is possible to read about this weapon in the popular science magazine *Gangut,* issue 14 for 1998, which sells for 40 rubles at military bookstores.

Articles that raised the possibility that the charges against me were bogus remained rare, however.

Much later, I learned the facts behind the rumors of my release. Sometime during the spring, Admiral Inman had called Ambassador Tom Pickering of the State Department on my behalf. Pickering was then the most distinguished senior career diplomat in the State Department, having risen through the ranks during both Republican and Democratic administrations to serve as ambassador to Nigeria, Israel, the Russian Federation, and the United Nations. He was currently in charge of all the American embassies and legations serving the countries of the former Soviet bloc, reporting directly to Secretary Madeleine Albright. At the request of Inman, Congressmen Peterson and Weldon, and others, that spring Pickering established an interagency team to look into my case and make recommendations for action. The team included senior officials from State and the Department of Defense, as well as from other agencies, and was coordinated by Andy Wise of the Russia country desk at State. The group met weekly to discuss strategy and exchange information.

It was this group's recommendations to Secretary Albright, the National Security Council, and to the president that formed the basis for President Clinton's involvement in my case and for the hope that he might obtain my release in June.

But the Bill Clinton who arrived in Moscow in early June 2000 was a weakened president heading into the last half year of his term in office. The Arkansas bar had recommended that he be disbarred; Clinton's heir, Vice President Gore, was trailing badly in the polls looking toward the November election and was distancing himself from his boss; the president's programs were stalled in Congress; and there was little progress in the Middle East peace talks that he was attempting to shepherd. Clinton had reportedly sought this meeting in Moscow, in the waning days of his administration, to try to create a foreign-policy triumph to match those of Reagan and Bush. He had in mind a "Grand Compromise" in which Russia would agree to allow the U.S. to build a limited antimissile defense system but not other new defenses in exchange for severe limits on both sides' numbers of nuclear warheads.

Prior to this summit the Clinton administration had done many things to placate and assist Putin. It had urged the World Bank to release $100 million that had been held up pending the rescheduling of Russia's debts. It had cut back on American sales of arms to Taiwan in an attempt to make less necessary Russia's renewal of its arms sales to China. Such actions spoke louder than the Clinton administration's words, for example, those of Secretary Albright in condemning Russian military excesses in Chechnya. On his way to Moscow, stopping in Germany to receive an award, the president suggested that Europe needed to work harder to include Russia in both NATO and in the European Union.

President Clinton's stated agenda at the summit was to persuade the Russians to accept changes to the 1972 ABM treaty, which would permit the U.S. to build a limited antimissile defense system. Republican presidential candidate George W. Bush had been campaigning on the notion of promoting a strong antimissile defense system; Bush had been accusing the Clinton-Gore administration of being locked into a Cold War mentality and had warned Clinton not to tie the hands of the next American president by negotiating a far-reaching treaty with the Russians just then.

But Putin evidently had no intention of negotiating with a lame-duck Clinton, and the June summit produced almost nothing of substance. During their visit, the Clinton delegation annoyed the Russian Federation president with two actions: Secretary of State Albright visited the offices of Radio Liberty, one of whose correspondents, after being held in Chechen and Russian prisons earlier in the year, had become an issue during the Putin election campaign; and Clinton himself went to the studios of the radio program *Echo Moskvy* at the flagship station of the Gusinsky media empire, where he put on headphones and answered questions from callers.*

Later I learned that Cheri and Keith, through my friends in the intelligence community and the offices of Congressmen Peterson and Weldon, had all tried long and hard to put my name on the agenda for the summit at some level, even at a lower level, such as the "bilateral" talks between Albright and Sergei Ivanov, but as far as they knew no formal mention of me had been made in Moscow. In fact, it appears that the subject of Ed Pope may well have been brought up at this June summit as a result of the Pickering-led group's work, but the negotiations between the U.S. and the Russian authorities had not yet reached the point at which the Russians would agree to release me. In any case, the summit passed, and I was not released. I guessed that if there had been negotiations, they had come to naught.

There was one tangible product of the Clinton-Putin summit. A few days afterward, on June 6, a man named Pavel Astakhov showed up at Lefortovo, and I met him in Shelkov's office. I should have been more attuned to the alarmed reaction Pavel provoked in Shelkov, but I was depressed at not having been released as a result of the summit, was still not sleeping more than a few hours a night, had received no letters from my family, had lost more than twenty pounds, and had become more than a little paranoid. Pavel's smooth appearance

*Most of the callers' questions were about Monica Lewinsky.

did nothing to disabuse me of my suspicions. He was a handsome, well-groomed man of thirty-four, on whose broad shoulders hung a tailored suit that almost certainly had been made for him abroad. He exuded an air of confidence, even arrogance. He spoke in cryptic terms, and in fairly good English. "You know who was here in the last four days, don't you?" he asked.

"Yes," I replied, assuming that he meant President Clinton and the U.S. delegation brought over for the summit.

"We were contacted by someone in the delegation, and asked to immediately begin to help you."

"Who in the delegation?"

"I can't tell you that."

But he could brag that he, Pavel, was very well known and had figured in many important cases named in a magazine article he showed me; but I didn't recognize the names he mentioned, and he didn't say Gusinsky's, although the MEDIA-MOST baron would shortly be Pavel's most-publicized client and I would have recognized that name had he dropped it. He also did not tell me something else that I would have found of prime interest — that he used to work with the KGB; had he given me that information just then, I probably would have dismissed him out of hand.

In the absence of a directive from my wife, I could not agree to have Astakhov represent me, no matter who had sent him. He understood that and told me he would be back when the arrangement had been properly explained to me. Later that day I asked Brad Johnson who Pavel was, and Brad said he didn't know; during his next visit, two weeks later, Brad was mortified at his mistake: Pavel Astakhov, he told me with awe in his voice, was one of the most celebrated defense attorneys in Russia, a frequently seen guest on television talk shows, a former KGB lawyer who was now hated by the security forces for actively defending the people they were trying to detain and convict. Brad opined that there wasn't a better lawyer in Russia and apologized for not having recognized Astakhov's name earlier.

I was still reluctant to accept Pavel Astakhov and his associate Andrey Andrusenko as attorneys until I'd heard directly from Cheri or Keith that it was okay to do so.

As it happened, on June 6 I was also given a present by my interrogator — not Blubber-Butt but the younger and friendlier Andre, who presented me with a small Soviet-era television set. This present, too, sent a mixed message: because although the TV was a gift, I was not permitted to keep it in my cell; it was put into storage with my other belongings. What a *durdom* (crazy house) this place was!

Whereas Blubber-Butt had meticulously gone over the "sensitive" documents, Andre examined my personal belongings. The contents of my wallet fascinated him. I was worried about the implications of one of the business cards. It wasn't the card from a woman at Lockheed-Martin; although Andre fussed over it, I could easily explain it as being from a person at a company whose business I had tried and so far failed to obtain. Rather, I wanted to protect the card given to me by a Russian from an institute in a city long off-limits to the West; the man was an important scientist who wanted to make a deal on some fairly sensitive though unclassified technology, which he had invited me to the institute to see. To divulge its origin would have precipitated questions that I didn't want to answer and likely would have gotten both the scientist and me into trouble. So when he asked about the card, I lied, saying that the scientist had been at a conference in the U.S. and had given me his card there. Andre accepted my explanation and moved on. Momentarily I wondered if he had done so because he recognized the scientist as an FSB operative — as I believed to be the case.

Andre spent an inordinate amount of time on three nonbusiness cards in my wallet: an ATM card issued by the Navy Federal Credit Union, a private bank catering primarily to U.S. Navy personnel around the world; my official Navy ID, with RETIRED written on it; and my membership card for the American Legion. To him these cards were highly significant, worthy of being added to the pile of "evidence" that marked me as a military spy. He spent an entire two-hour session trying to get me to agree to his contention that although my ID card from the Navy said, in plain letters regarding duty status, RETIRED, the label was a way of disguising that I was actually still on active duty. To support his contention, he deliberately misread the date of issue on the card, he questioned serial numbers and birth dates, and he insisted that the background pattern on the card contained coded messages. In other circumstances, Andre's just plain stupid attitude might have been funny, but as I sat in his interrogation office it made me despair. His mind had been made up, and he had no intention of being confused by the facts. How would I ever be able to prove my innocence?

CHAPTER 5

Explosives, Exclusives, and Late-Night Visitors

Many people have opined that my arrest was inevitable, given the climate of doing business in Russia. Most of them have never been to Russia, and I don't agree with their Monday morning quarterbacking. Before my arrest I had been working in Russia for eight years, making business transactions with its scientific institutes and manufacturing entities; various Russians had offered me technologies far more sensitive than the one I was accused of stealing — and I had refused them — and some Western companies had requested that I pursue far more secretive products than the propulsion system of an old torpedo, and I had refused those requests, too. But I will agree that doing business in Russia is like balancing on a high wire: the risks are significant and not always in proportion to the potential rewards.

As the Soviet Union was falling, in 1991–1992, the entire panoply of Russian defense technologies came on the market. "It is an amazing scenario," the *International Defense Review* observed:

> The doors of virtually every former Soviet defense-science laboratory [have been] thrown open to the West. Now the defense technologies of the Soviet empire . . . are on sale to governments, industry and academia [in] a high-tech, Wild West free-for-all

> with the creation of dozens of new exchange institutes, technology "brokers," and business speculators, on both sides of the former Iron Curtain, all scrambling to facilitate deals.

I found a quote from former French president Georges Pompidou that after a few visits to Russia I had framed and put over my desk at ONR: "There are three ways of going broke: gambling, women, and implementing high technologies. The first is the quick way, the second is pleasant, and the third is reliable." To prevent the reliable from happening to me personally and to the American organizations I represented — at first the Navy, then private enterprises — I developed a series of rules for doing business in post-Soviet Russia.

The first rule: Never completely trust a Russian. The former Soviet people are very much unlike us, and it's not just a matter of style. When dealing with the Japanese, you must be mindful of cultural differences but you work within the same framework of buying and selling as we do. Not in Russia. In many ways Russia today resembles the American Wild West of the nineteenth century: a lawless, every-man-for-himself, continually perilous arena. In Russia there are no assurances about the quality of merchandise (unless the product has been produced for the military), delivery dates, or reliability; no guarantees of origin of materials; and no conception of property rights, valid contracts, proper pricing, or commitment to a partnership.

What else could you expect from a country that before suddenly becoming a capitalist society had been a Communist dictatorship for seventy years and whose leaders had continually fed its people lies? The Soviet government isolated the Russian people from the rest of the world, taught them that the West was corrupt and evil and that the worst elements of the West were embodied in capitalist businessmen, regularly depicted as ogres who cheat the poor, take advantage of workers, and destroy the weak and helpless in their insatiable quest for plunder. Thus, Russians actually came to believe that Wild West capitalism — a crooked, entrepreneurial, dog-eat-dog form — was the norm for a capitalist society. When they suddenly found themselves in their own free-market society and were expected to act as capitalists, they thought they were obligated to behave like ogres. They started doing "biz-ness," as they wrote and pronounced it, without having a clue as to what the word or the practice meant.

Russians are not evil to their core; I like them as individuals and admire their cultural traditions and the quality of their scientific insights and ingenuity. The people I dealt with in universities and institutes spent most of their lives employed and protected by the state, with guaranteed salaries and lifelong benefits. After the fall of the Soviet Union they were still in the same lines of work but no longer had a guaranteed salary or a guaranteed job. This made them desperate and willing to do almost anything to survive. A quote from a former nuclear physicist, printed in an article in the *Moscow Times,* well illustrates their attitude:

> I'm not interested in politics, because I don't care which clerk I give my bribe to. Before, I was involved in what eight to ten other people in the world do, but any academic boss could "smear me on the wall." And now I can show the government that I'm smarter than its entire machine, because I know how to get around my new restrictions. I derive pleasure first and foremost from the recognition of my own intellectual superiority.*

Russians regularly lie, cheat, and steal in business — all Russians, including heads of prestigious scientific institutes, world-renowned scholars, and my friends and business partners. Honesty, truthfulness, fair and open dealings with Western businessmen and with one another, are not merely a luxury they cannot afford; to Russians these are unfamiliar and ineffectual business practices. For instance — and it pains me to write this — I knew that my partner Bolshov, although otherwise trustworthy, overcharged me for transportation, translation, office, and travel services. I accepted this as necessary for

*Quoted in Tatyana Matsuk, "Honest Business," *Moscow Times,* March 17, 2000.

him but wished he would simply tell me that his costs of doing business, his need to succor his extended family, and the uncertainty of the world in which he lived mandated his trying to obtain from me every last *kopek* — but he was unable to do so. For people such as Bolshov, the many years of life in a police state had convinced him that deceit was the key to survival, so being aboveboard with a partner in the Western sense was impossible.

Bolshov recognized that I wasn't a millionaire. But many Russians acted as though I were — as Soviet propaganda insisted that all American businessmen were. Russians who were highly sophisticated in other matters actually expected me to come over to their country, sign a contract, and, right then and there, deliver to them a check for millions of dollars. Frequently one or another group of Russians would present a proposal with a price tag in the millions; I'd tell them that if their idea was good, we'd start with a small development grant in the tens of thousands of dollars. "Okay," they'd instantly chirp, "we can work for that." At one institute the director initially sought my assent to a $1.5 million deal for a research project that he and his staff eventually agreed to conduct for $75,000.* After I'd shaken hands on the latter amount, that evening there was a knock on my hotel door: it was an engineer from the plant, telling me he'd get the same research done for $5,000. "*Nyet problem,*" he said of that amount. I mournfully informed him that I would have to stick with my arrangement with the institute's director. During eight years of traveling to Russia, there were many similar late-evening knocks on my hotel-room door from men trying to sell me any technological thing that could be sold — and plenty that could not be. A chemist from Ukraine was one such late-night hotel visitor; he brought a sample of an underwater adhesive that his institute had developed for the KGB. It was to be used to attach a limpet mine to

*Another program that we negotiated carried an initial asking price of $100 million; when we proposed $50,000, the response was a delighted *da* and a demand for a 50 percent prepayment.

the underside of a ship: saboteurs living abroad would keep the adhesive and mines handy, and in case of impending war would receive instructions to put them to use. I understood that there were dozens of potential commercial uses for the adhesive — for instance, to repair undersea cables. But I couldn't consider buying it from this individual, because he almost certainly had no right to sell it. He professed astonishment that something developed by the KGB was of no interest to me.

Frequently I'd be presented with a contract, not because the Russians wanted a contract but because they believed Westerners needed contracts to consummate deals. We do need contracts, but not the kind they had in mind. Russian contracts are not the same as ours; the Russian legal infrastructure is weak, whereas the American system is bursting with lawyers who know how to write contracts that meaningfully bind both sides, and courts that have the power to enforce penalties if one or the other party does not live up to the provisions. In Russia a contract is simply a platform for extracting a 50 percent down payment from a foreigner and for staging further negotiations to raise the price of what the contractor *may* eventually deliver or produce.

Similarly, Russians would offer me an "exclusive" on a product or process; they knew Americans liked exclusive deals, so one would be suggested even if an exclusive had already been sold to five other buyers. A company in California called me at ONR to boast that it had bought the entire Krylov Shipbuilding Research Institute and had an exclusive on its operations. This was like admitting they had been sold the Brooklyn Bridge; Krylov was a vast place with 10,000 employees and could not have been sold in its entirety. All the Californians had bought was the right to work with one breakaway design team of four engineers.

Again and again, Russians offered me products and processes developed for the military that I knew would get the sellers and me into trouble were I to buy them. In St. Petersburg the proffer was for a high-frequency acoustical-optical sensor that had clearly been developed for the most advanced Russian submarines; the potential seller's line to me was "our Navy won't buy it, but yours may." On behalf of my government, I politely said no. I turned down an offer from a junior engineer admiral on active duty with the Soviet navy, as well as similar ones from KGB and GRU officers, all of whom said that if I would get them jobs in the U.S. they'd give me half of their first year's salary.

I was regularly offered weapons and explosives. While visiting one institute where my primary interest was in an optical coating for a lens, a scientist in an attached lab insisted on demonstrating for me a new explosive. A pinch of it did an awful lot of damage, and I was certainly impressed; but I told him and his director that I didn't want to see things that were obviously still military secrets. At a shipyard in Nizhny-Novgorod I was shown a 5,000-ton section of a submarine hull, which had been removed from a construction bay, and was told I could buy it for a song. It was made of titanium; at that point the Russians referred to me as Captain Titanium because of my interest in the metal, and they went out of their way to try to sell me titanium in any form — even though its export was illegal and required special permits. "Of course we can obtain permit for you."

While still on active duty, I attended a meeting at a Russian institute whose director wanted our Navy to cooperate with the Russian navy in building a low-frequency, shallow-water submarine detector. As I listened to this astounding proposal — a cooperative U.S. and Russian venture to mutually guard our countries' coastal waters against intrusions from third countries' submarines — I thought the KGB man in attendance might get up and shoot all of us for even conceiving of such a thing. He didn't; at the end of the meeting, he said that it was a wonderful idea and that the KGB would help if it could — which I took as meaning that if we paid the KGB directly, there would be no problem. That I would not do. Nothing further was heard of the matter.

The second rule of doing business in Russia is not to restrict yourself to interacting with one individual or institute. Russians want to monopolize your time, take you under their wing, not just sell you their own product or service but be the intermediary who obtains for you every other product imaginable. Instead of passing you on to others, they'll stay with you, take you sightseeing, and invite you to their homes, sometimes to be friendly but more often to prevent you from doing business with anyone else. "I can do best work for you," a typical partner would say. "Everyone else is bum."

I was warned against certain individuals and institutes known for greed and nondelivery. One that was involved with plastics and polymers wrangled a $250,000 contract from a large and prestigious American firm; the director of the institute used the money for a new apartment, a BMW, and a yacht. When the buyers asked for the products they had ordered, they were told that the project had run out of money and that another $100,000 was needed to produce anything.

The directorship of an institute was awarded to a man I'd come to know, so I went to his new office to congratulate him. He was called out of the room for a time and then returned, ashen. The Mafia had just told him that his predecessor had been paying them for protection and warned him that if he wanted to stay alive, he must do the same. On my next visit there were guards outside the institute whom my friend identified as Mafia men that he had been forced to hire; these guards drove Ford Explorers and BMWs, while the institute personnel could afford only broken-down Volgas. My friend gave me a pennant from the new security company; it featured a version of the KGB shield, but with a Mafia-related name, and its display at the gates of the company would have warned Russians, who knew that shield all too well, that the enterprise was linked to the KGB and the Mafia. I keep the pennant on the wall of my office as a reminder of that connection.

Shakedowns were routine operating procedure all the way to the top. I learned of an incident involving a diplomat from North Korea who had tried to reenter Moscow with several million dollars in counterfeit American currency and heroin worth almost as much. A drug-sniffing dog pointed to the diplomat's luggage with such excitement that the airport guards found the contraband. But just as suddenly, a higher-up in the FSB organization interceded, and the diplomat proceeded to his embassy with his bags. The explanation given to me was that the FSB was in collusion with the emissary in the distribution of both the counterfeit dollars and the drugs. Many similar stories demonstrated the interpenetration of the KGB/FSB and the Mafia, working together to keep the Russian population terrified, poor, and dependent.

In Russia all information must be bought, and you can trust only the information that you have properly paid for. After making a few wrong guesses, I decided to deal solely with heads of institutes and to insist that they show me documentation proving they were legally empowered to sell me what they were offering. Intellectual-property rights? They'd never heard of 'em. Patents? Very few patents on technology were registered. There were "author's certificates" for technologies originally sponsored by the military, but if I asked for an author's certificate for a particular technology, I'd be told either that it was unavailable or that it was still classified.

At times I was shown certification papers that had SECRET stamped on them and a handwritten *X* crossing out the classification. "I can't accept these," I'd tell my hosts. They'd confer for a few minutes; then one of them would go into a back room and make a photocopy of the certification papers while holding a blank sheet over the classification logo. Or he'd take a razor, excise the stamped portion, and then present the copy to me as though nothing had been changed on it.

Yet things that should have been kept secret — very secret — were readily available and completely legal to buy. At a commercial model shop, I made an astonishing purchase: a one-of-a-kind model of a Russian submarine, of a class of subs that had not yet been launched. It was called the *Yuri Dulgaruki,* after the founder of the city of Moscow; I deduced that although the scale model had been made, production of the subs themselves had been slowed down by the economic chaos — and that someone from the Rubin Design Bureau had spirited away this model and sold it to the shop. I bought it for $420, kept it in my hotel room in Moscow, then took it home.

For a project or a purchase that isn't being completed right away, Russians want you to proffer a binder or a down payment; six months later the binder money has disappeared, but nothing else has happened. If the finished product does eventually materialize, it will not come into your hands for the price you first agreed to pay. There'll be add-ons. Delivery charges. Export charges. Handling charges. Extra charges that you know will be going to the FSB and the Mafia. Negotiations will continue until you give up and pay what is demanded, or walk away. To counter such practices requires a thick hide and plenty of skepticism.

And an ability to drink vodka.

An American walks into a business office in the U.S., looks at the product line, asks germane questions — about the product's characteristics, reliability, price, etc. — and makes a deal. Not in Russia. Upon arriving at an institute or a business entity, you have to get acquainted, engage in small talk, have tea and cookies. Then, if that has been successful, vodka — sometimes quite a bit of it, and no sipping allowed. If you don't drink with us, you're not a friend. Then it's off to tour the factory floor or the museum or the design bureau or the production facilities. I would visit facilities that had once been top-of-the-line research and production places; machines would be corroding, supplies gathering dust, factory floors empty and depopulated but for a handful of technicians in a corner doing freelance work with the factory's machinery. "Give us this contract," I'd hear, "and we can staff up quickly."

Once, to prove their worthiness for a research contract, officials from a St. Petersburg institute took me forty miles outside of town to a facility that had previously been off-limits to Westerners and showed me their museum of undersea technology, which contained quite a few devices and technologies that I recognized as having once been top secret; I wasn't allowed to take photographs but was given a glimpse of their treasure as a demonstration of their bona-fides.

After the factory tour, perhaps a day or two later, Russians are ready to do business and begin negotiations, which never conclude on the first go-around. It would take me two or three trips to Russia, each with a visit to the same facility, to finalize a project. When you find the right institute to sell you the technology, then the people with the proper set of skills have to be identified and located. They are found, but the appropriate machinery is unavailable. After all the elements have been put together, you learn that the institute is regretfully unable to procure the right raw materials for the job. Or there was a fire. Or the institute director's mother died. Getting anything done in Russia is never simple.

On a typical, three-week visit to Russia in August and September 1998, the following events occurred. At a small R&D company, I was pursuing a new way of producing low-carbon steel; during conversations with management, it became apparent that only one manager was potentially trustworthy — all the others were suspicious and naive — so it would be difficult to complete transactions. Another company had a novel cooling technique for cutting and lathing operations that I had been attempting to license for two years; during this visit I had two meetings with the directors. They gave me a videotape but told me that if a U.S. purchaser was really interested, TechSource should take the Russian team to the U.S. for a demonstration. I agreed to do so if they would give us an exclusive license. They said yes, then refused to sign such a license, saying their word should be sufficient, and tried to sell me a nine-year-old instrument as a demonstrator, at a ludicrously high price. I walked away. Later, at a third institute, I looked at a device for cleaning diesel engines and was put off by its high price relative to similar products available from non-Russian sources; it was explained to me that the Soviets had never put a premium on price, only on making the device, which resulted in the production of devices that were more complicated and more expensive than they needed to be.

I did eventually manage to buy a fascinating technology from the Krylov Shipbuilding Research Institute: originally developed for ships, it used an electric current to counter the electrolytic reaction between metal and seawater that can result in the corrosion and failure of metal hulls. I had a U.S. client who wanted to counter the similar corrosion affecting underwater cables that carried electrical current from Connecticut to Long Island. Those cables had cost about $60 million to lay, and the tide and sea currents were promoting electrolysis on the surface of the cables, degrading the coating. The Krylov system would measure the temperature and salinity of the water and the speed of the current, then automatically send out the right amount of electric current to prevent or retard the electrolysis. The potential savings to my client would be huge, maybe $200 million in delayed replacement costs. A Russian whom I had come to know wanted to sell this system with the cooperation of the Krylov people, but there was too much red tape: the Russian government wasn't eager to sell it to the U.S. but was willing to sell it to other countries. So he set up an office in Finland, brought it there, and resold it with the help of my company to the eventual user in the United States. There is no doubt in my mind that he was able to complete the deal only because he was kicking back a percentage of his share to individual engineers and scientists at Krylov.

Occasionally I'd be able to complete a simple and straightforward transaction for a unique and extremely useful technology, for instance, an "acousto-optical tunable filter" spectrometer, which measures wavelengths of light. It was designed by Vladislav Pustovoit, originally for use in detecting the paths of submarines through analysis of the pattern of water disturbance in the wake of a submerged or surface boat. The filter was a real gem, and I was amazed that a technology so good and with so many potential nonmilitary uses would be permitted out of Russia — but it was for sale, with all the proper papers attached, and I bought it legitimately and shipped it home. Pustovoit's ingenious device was built around a solid crystal; because it had no moving parts, it was perfect for missions into space, where it could perform such tasks as measuring plant growth. I thought it could also be used from space to measure pollution on Earth in new ways, and I had other applications planned. I kept it at home and was lining up many potential buyers for it in the U.S. When the Russians arrested me in April of 2000, this was one of the devices they initially accused me of stealing.

Many observers likened the opening of Russia to the discovery of gold, but I preferred to think of Russia as a gigantic flea market in which everything was for sale, but to obtain the best-quality merchandise, you had to search for the real treasures, cut through the chaos, deal with all the cheating, and bargain adroitly.

My mistake was to believe that I had the Russian rules and roadblocks all figured out; I had not factored into my calculations that the basic ground rules might suddenly and drastically be changed — without notice — and that this change would result in my arrest and incarceration.

CHAPTER 6

Incidents in Limbo

The concept of limbo derives from a theological notion: being between heaven and hell, unable to enter either; the dictionary defines limbo as "a place or condition of confinement, neglect, or oblivion." For me, Lefortovo was limbo. My cell mates and I were consigned there not as prisoners but as detainees. The uncertainty of what might happen to us, and when it might happen, weighed on us heavily and constantly. Some detainees at Lefortovo had been in true oblivion, held there for up to three years without charges being filed or dates for trials being set.* No magistrate had the power to independently inquire why someone had been arrested. No lawyer could arrange for a detainee to be released on bail or obtain from a writ of habeas corpus that would force the FSB to bring the case to trial or dismiss it. A detainee was deemed guilty until proved innocent; accordingly, the FSB rules for holding us were designed to exacerbate our uneasiness and hurt. For instance, you could receive a letter but could keep each letter only one day, and at ten that night it was taken back by the guards and stored with your belongings, out of sight. What reflexive cruelty!

*Russian law provides that the limit for pretrial detention is eighteen months. Several prisoners whom I met, and others I heard about, had been detained for much longer. And one cell mate claimed that a friend had been "bought out" of a Russian prison for a payment to the FSB of $100,000.

It was compounded by never knowing whether all the letters sent to you from outside were being forwarded to the cell. Brad Johnson reported to the State Department that he had delivered to the FSB sixty-seven letters for me, but the system refused to pass on most of them. Similarly, with funds from Cheri, Brad had bought fruit and other food and conveyed it to the prison, but only some of the food had been delivered to me — and that only upon my specific written request for an item. Boxes of crackers and cookies sent by outsiders would be opened by the guards, the contents examined and then emptied into a plastic bag, mingling the salty and the sweet in a way that made a mockery of the gift. You could not make or receive phone calls. In the exercise rooms you could not speak English or shout, to prevent you from learning whether you had compatriots in other cells. In one bag from the consulate, there were envelopes and postage, but no writing paper, so I wrote my wife and family on brown toilet paper just to show them that I had not lost my sense of humor.

When the Clinton-Putin summit in June did not bring about my release, I experienced periods of anger and depression. For Cheri and her advisers back home, however, this moment was much more of a turning point. Since there was no movement toward my release after the June summit, they decided there had to be a change in strategy.

By mail, Federal Express, e-mail, and every other means, Cheri had been sending requests for help to various agencies and departments in the White House, to the State Department, to the Department of Defense, to the vice president and every cabinet member, and to a number of Congressional Representatives and Senators, once or twice a week. In partial response, Mark Medish of the National Security Council had phoned her to say that the White House was doing what it could to get me out of prison. Despite that assurance, I was still in limbo. "The family and I are devastated by Clinton's lack of addressing Ed's situation and we cannot in good conscience let this go any longer without public awareness," Cheri explained in an e-mail to a friend. Now Cheri determined to make something happen — by going public.

It was a decision with which Congressman Peterson and his staff concurred. They already had a high-placed and responsive ally in the State Department, Tom Pickering, who along with Scowcroft and my former commanders in Naval Intelligence (Brooks, Studeman, and Inman) continued to apply behind-the-scenes pressure to free me. But the plain fact was that the quiet-diplomacy effort had yet to produce any visible results.

Many of my supporters could not understand why State and the White House were reluctant to apply enough pressure to the Russian Federation to spring me. Some believed the reluctance stemmed from President Clinton's long-standing antipathy toward the military services; others attributed it to Clinton's and Albright's apparent belief that the American president could not ask Putin to release me and at the same time ask the Russian Federation to yield to the U.S. on the anti–ballistic missile shield agreement. Both groups of my supporters scoffed at State Department bureaucrats who argued that any attempt to pressure the Russians for my release would be construed as inappropriate meddling in Russia's judicial system. Such an attitude might be appropriate if this were a drug-smuggling case, but since espionage involves giving secret materials to a foreign government and since the American government knew I had not forwarded any secrets to it, State would not be meddling if it interceded in this case; rather, the U.S. could, in effect, assure the Russians that the alleged crime had not even taken place.

To change the attitude at State and the White House, Cheri and my friends decided to use the only weapon in their arsenal to which politicians and politically appointed bureaucrats are always sensitive: the media.

In the previous months Cheri had received dozens and dozens of calls from the press and had refused to be interviewed or give statements; in contrast to the experience of many who had been suddenly thrust into the limelight and who came to view the media as a pack of sharks in a feeding frenzy, Cheri found reporters to be generally sympathetic, willing to understand why she could not say anything publicly, and respectful of her privacy and of the family's pain. Now she determined to hold a press conference, in part to respond to the media calls that she had earlier deflected. This was a huge decision for her, not only because it contravened the previous quiet-diplomacy strategy but also because it would thrust her into a public role; she was uncomfortable with such a role and frightened that in filling it, she might somehow make a mess and inadvertently damage me.

The press conference was scheduled for June 8, until Congressman Weldon asked to postpone it so that he could make a direct appeal while in Moscow that week, accompanying Secretary of Defense William Cohen. During his official visit to Moscow, Weldon repeatedly tried to see me at Lefortovo but was rebuffed by the FSB. He also got nowhere with a similar request made to the second-in-command at the Ministry of Foreign Affairs. Secretary Cohen reportedly brought up my case with his opposite number, the minister of defense, who also took no action.

In Moscow, Weldon showed the press what he had offered to Russian officials, documents that Keith and Cheri had gathered: copies of my business contracts with the various Russian institutes and the official Russian endorsements of what I had been doing. Weldon reminded the Duma representatives that he was the best friend of Russia in the American Congress and that a $25 million aid-to-Russia bill, then before Congress, might be compromised if I was not released soon. There was no official response.

The first salvo in the public campaign on my behalf began with an article written by reporter Dave Montgomery, who had been with Weldon in Moscow and had interviewed Cheri and others working for my release. It was printed in the *Philadelphia Inquirer* and other Knight Ridder papers. It made the points that my rare cancer was in remission but had to be monitored and that I had also recently been diagnosed with Graves' disease, a dysfunction of the thyroid gland.

The day before Cheri's rescheduled press conference, the FSB arrested Vladimir Gusinsky and held him for several days. Pavel Astakhov received a lot of airtime arguing in public and in court for

Gusinsky. A high-level U.S. delegation led by a former ambassador to the U.S.S.R. canceled its visit because of Gusinsky's arrest. Fifty congressmen protested it. President Bill Clinton not only telephoned President Putin and asked him to reconsider the jailing of the MEDIA-MOST magnate but also leaked word to the American press that he had done so. Putin then made a public statement implying that Gusinsky's arrest was unjustified, and the media baron was released.

President Clinton could put in a good word for the Russian oligarch whose broadcasting facilities he had used but could make no mention to Putin about Ed Pope, an American who had spent most of his life in the service of his country? Clinton's apparently misplaced concern incensed my family and friends; had I known about it then, in prison, it would have enraged me, too.

At the press conference on June 15, Cheri read a prepared statement that stressed my innocence, the suffering of my family, her fears that I was physically deteriorating in Lefortovo, and her fervent plea that the U.S. government do more to obtain my release. Her remarks were followed by those of Congressmen Peterson and Weldon; Peterson spoke of having sent a letter to the chairman of the House subcommittee that oversees the foreign aid budget, recommending that funding specific to Russia be held up until I had been returned home. At his urging, other congressmen sent similar letters. He also pledged to examine areas of the military's aid budget in order to similarly apply pressure on Russia.

As a follow-up, Fox News's Teri Schultz (a friend of Jen Bennett, Peterson's press secretary), whose regular beat was the State Department, interviewed Cheri for Brit Hume's program. "Cheri was very nervous, of course," Bennett later recalled, "but she came through as completely genuine and believable." Also, at a routine State Department briefing, in response to questions from Schultz, spokesman Richard Boucher for the first time provided the press with concrete reports on my condition, based on information from Brad Johnson; previously, reporters' questions about me had been answered with a "We'll get back to you on that." From this day forward, Jen Bennett would regularly suggest certain questions for Schultz to ask about my case, and soon the competitiveness of the reporters on the State Department beat kicked in and they began to ask similar questions about me, which Boucher became obliged to answer with greater specificity and detail than State had been used to providing. "Did you know that Ed was denied a doctor's visit today?" a reporter would ask in a daily briefing, and Boucher would have to obtain up-to-the-minute information with which to make a proper response — or risk appearing foolishly uninformed.

Days after the press conference, Cheri and two senior Peterson staffers, Bennett and Bob Ferguson, flew to Moscow. When her earlier applications for a visa had been snarled, Cheri had threatened to go to Sweden and enlist the help of friends there to enter Russia carrying Swedish papers, if necessary. Her determination to visit me at any cost was one of the factors that had pushed Peterson's office to ask their high-level contacts at State to make the proper arrangements, though Peterson himself was unable to accompany Cheri because Congress was in session.

On hearing from Brad that Cheri would be coming to see me soon, I went into a disciplined frenzy of work, compiling pages of notes on subjects I wanted to discuss — really, had to discuss — with her: the location of bank books, insurance policies, my medical records, and other documents that had become extremely relevant. Every emotion possible went through my head at the thought of this long-awaited reunion: anticipation, dread, longing, fear of something going wrong and of my being too emotional to say what needed to be said, and guilt at my having put her through the great difficulties that, without knowing the details, I understood she had been encountering.

Even with high-powered shepherds from a congressman's office, Cheri and her party received incredibly shoddy treatment from the

American embassy in Moscow. "They refused us transportation and laughed at the idea of providing security," Jen Bennett later recalled. She did manage to get Brad Johnson to agree to meet them at the embassy, but he told them to go to the old building rather than to the new one. Cheri, Ferguson, and Bennett were kept waiting outside for two hours in a line reserved for Russians attempting to obtain visas to come to the U.S. When Cheri's hired security men tried to hurry things up, the Russian guards on the street outside the embassy were hostile. Inside, waiting with half a dozen others in an antechamber, Cheri was shoved by another Russian guard; when she pushed back, an American Marine yelled at her not to touch the Russian and warned that if she did, he would throw her out of the embassy.* More delays followed. Cheri could not help but feel that the indignities were deliberate, an impression that became solidified when she realized that the man behind a glass partition, watching her with the Marine guards, was Brad Johnson. And she did not think that the impetus for the indignities came from Johnson; like the Marine guards, he was just following orders.

Eventually Johnson brought Cheri, Ferguson, and Bennett to meet with more senior embassy staff, though not with Ambassador James Collins. For three months State had refused to say categorically that Ed Pope had not been a spy, even though by this time they had been assured that this was true by the CIA, the DIA, and the Navy. State's outright refusal to say publicly that I had not committed espionage contributed to the Russians' impression that I was guilty. While in Moscow, Cheri asked again for the embassy to put out a statement that Ed Pope was not guilty of espionage, and embassy officials told her they could not do that. Brad Johnson was also reluctant to provide Cheri with the list of lawyers he had presented to me — he said he believed she might turn it over to the media, who would quickly

*During my twenty-five years in the Navy, I worked with many Marines and know how disciplined they are; they follow orders unquestioningly. No Marine would have treated Cheri as this one did unless he had been ordered to do so.

figure out that it had been taken straight out of the Moscow telephone directory — or a list of which letters and what food he had given the Lefortovo authorities to forward to me, but eventually he did so. And he readily agreed to accompany Cheri to the prison the next day, since the authorities there had said she could not visit me unless a consular official brought her in.

On Sunday evening the prison authorities handed me thirty-five letters that they had been holding for some time. Reading these letters was an overwhelming emotional experience — as I am certain the FSB had hoped, so that the letters would unnerve me for the next day's meeting with Cheri. I gobbled the letters like candy. There were missives from my sister, Brenda, in Denver, from other family members and friends, from people I had known years ago and had all but forgotten, and from people who did not know me at all but wanted me to know that I was in their prayers. There was also a gag letter from Thumper, our little white poodle.

On Monday morning I was escorted into the VIP visitors' lounge at Lefortovo and told to wait, along with guards. It was an ornate place with Italian furniture and Oriental rugs, the most lavish location I had seen within the prison. Cheri was ushered in.

My wife of thirty years was a wonderful sight for sore eyes. But she looked terrible, her face puffy, eyes encircled with dark rings, worry and suffering evident in her visage and body language. Conflicting emotions raced through me: love, devotion, guilt — above all, tremendous guilt: I should never have gone to Russia, I should have come home when I became suspicious after the first meeting; Cheri had even advised me to do so when I'd apprised her of those suspicions in an early phone call. Now I didn't know whether she would reproach me or if she would even touch me. Maybe she'd treat me like a leper.

In an instant all my doubts were wiped away as she embraced me, telling me how much she loved me and how she was trying to help me. We both began to cry. I told her I loved her, adding, "I did nothing wrong. I'm not a spy. These charges are false."

I felt compelled to say these things — right away — for I didn't know how long our meeting would last. I also did not know what lies she'd been told by the Russians, what misinformation she might have read in the papers, or what the Navy or Penn State had said to her. For all I knew, she might harbor a lingering suspicion that I *had* been in Russia as a spy; after all, there had been so many years of our life together when I had not been able to tell her precisely what I was doing, although she knew it had dealt with sensitive intelligence operations.

Now she assured me that she knew I wasn't guilty of espionage or of anything else illegal. I began to calm down. I noticed that her clothes were nice, some of them new; they were for the public appearances and interviews she would be giving. We spoke aloud of family — of my father, who was ailing, and of her mother. Cheri told me they were both okay. That wasn't so. As I learned much later, Cheri's mother had died the previous Friday, and Cheri had been informed just before she boarded the plane to Moscow. Mother and daughter had been close, but when Cheri heard the news, she had said to Jen Bennett and Brad Mooney: "I can't help my mother anymore, but my husband is in prison and I can still help him. I'm going to Russia." Cheri recalled that in one of their last phone conversations, her mother had told her to be strong and to bring me home. She asked her family to put off her mother's funeral until her return to the U.S. On the plane Cheri felt her mother's presence by her, aiding her from afar.

Not wanting to burden me in our prison meeting, Cheri also didn't tell me of the difficulties she had encountered in obtaining a visa and in dealing with the State Department and the embassy in Moscow. For my part, not wanting to burden her, I did not mention the precariousness of my health, the fear I felt, or my suspicion that my cell mates, probably KGB plants, might murder me.

We held hands on the sofa and stuck to innocuous subjects. It had been three months since I'd conversed easily in English. Did you pay this bill? How is the National Governors Association conference planning going? Have you been getting exercise? There were plenty of matters we knew enough not to discuss, and I was determined not to reveal anything that might help these bastards hurt my family or me. The guards would not let me give her my pages of notes or the letters I had not been permitted to send. And she was not allowed to pass anything to me. Cheri told me that she had met Pavel and Andrey, who were to be my new lawyers, but did not explain any further. I informed her I'd read the Bible; she instructed me to read it again, because I might have missed something.

The worst moment of the day, I whispered to her, was at ten in the evening, when we were ordered to lie down and sleep, with the lights on. She vowed to share that moment with me, every day, by halting whatever she was doing and being quiet, half a world away.

The guards interrupted. The visit had to end, and there would be no more visits from her for at least a month. We kissed and separated — a very difficult separation for both of us. I didn't know whether I'd ever see and hold her again.

I saw her that evening but only from a distance, on Russian television, being interviewed on the street. She looked better — calm, collected, and soft-voiced, strongly insisting that I was innocent, saying she would not be dissuaded from doing everything in her power to bring me home. She told the interviewers that I had lost weight, seemed nervous and strained, and had an unexplained rash but that I still looked beautiful to her. The interview was also carried in the United States on Brit Hume's program and was the basis for stories from several news services.

I was enthralled by Cheri's strength and courage. The cell mates who watched the interview with me declared themselves in love with her, marveling at her loveliness and devotion. Her actions provided a stark contrast to the behavior of their own wives or girlfriends. In Russia when a man was arrested, his mate would typically shrug her shoulders, tell him to call her when he was released, and then try to find some new way to feed and shelter herself and her family. She would not fight the authorities for her man, because she would

Aboard the U.S.S. *Oriskany* during operations in the Gulf of Tonkin. As an air controller, I mistakenly vectored two planes over Hainan Island; fortunately, they exited Chinese airspace before an incident could occur.

The guided missile cruiser U.S.S. *Albany*; aboard her, I helped brief Prince Rainier about Soviet forces in the Mediterranean in 1979.

With a Romanian Navy officer and his wife, during our official visit to that country. On another frame of this roll of film, I confirmed the existence of some MIGs sold to Romania by the U.S.S.R.

The Whiskey-on-the-Rocks incident, involving a Soviet Whiskey-class submarine, SS-137, that carried nuclear-tipped torpedoes. It provoked an international incident. Behind the scenes, as a naval attaché in Sweden I provided assistance to friends in the Swedish Navy.

Participating in a "war game" at the U.S. Naval War College, Newport, R.I., 1986, during my years at the Pentagon working on advanced weapons and intelligence-gathering systems.

This is the "Red October" ship construction yard in the formerly closed city of Gorky, now Nizny Novgorod.

With my hosts during a visit to a large turbine plant in Kaluga. Guess which Russian is the FSB security officer.

Four major players in my story. From left to right, Professor Anatoly Babkin, later accused with me of espionage; Arsenty Myandin and Genrik Uvarov, who testified at the trial; and Dan Kiely of Penn State Applied Materials Research Laboratory, in whose briefcase and computer were found files that became the basis for my being charged with espionage. I took this photograph in the summer of 1996.

The Sayani Hotel in Moscow. Where I was arrested.

Lubyanka, KGB headquarters.

Statue of Felix Dzerzhinsky, once in front of Lubyanka. Toppled and trashed in 1991, it has since been mostly restored.

Lefortovo Detention Center. My prison for 253 days.

(AP/WIDE WORLD PHOTOS)

On the way to trial, August 2000. The first time the public had a glimpse of me since my arrest in April. (AP/WIDE WORLD PHOTOS)

The center of attention: the Shkval-E underwater high-speed rocket torpedo, as shown in a Russian advertising brochure widely distributed in 1998.

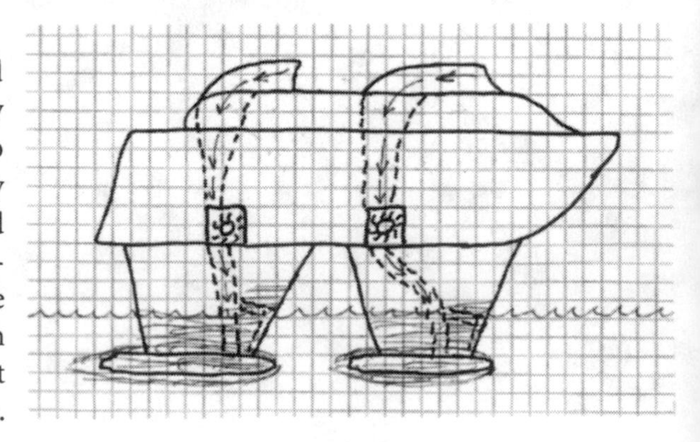

Drawing prepared by me during my interrogation to explain how my company planned to adapt the propulsion system of the Shkval for use with surface transport vehicles.

In the Defendant's Cage in room 227 of the Moscow City Court, during my trial. I had never felt so alone as I did in that cage.
(AP/WIDE WORLD PHOTOS)

Pavel Astakhov, my lead attorney, doing what he did so effectively: talking to the press, making my case in the court of public opinion.
(AP/WIDE WORLD PHOTOS)

Pennsylvania Representative John Peterson greets Cheri and me on the tarmac at Ramstein Air Force Base, Germany. My first steps on free ground.

Home again in Grants Pass, Oregon, being hugged by my mother.
(AP/WIDE WORLD PHOTOS)

As Grand Marshal of the 2001 Memorial Day Parade in my boyhood hometown of Grants Pass, Oregon, I proclaimed my position in the words also used on my new Pennsylvania license plate.
(TIMOTHY BULLARD/ GRANTS PASS DAILY COURIER)

For more photos, visit www.edmondpope.com.

accomplish nothing by doing so — except, perhaps, to get herself in trouble, too.

Cheri's extreme "Stand by Your Man" stance and her fervor and determination on my behalf were more than admirable, they were essential to me, giving me hope for continued life. Now, more than ever, Cheri was my heroine.

As the *Moscow Times* put it, "When the news broke in 2000, there was one man almost invariably on the scene," Pavel Astakhov:

> Often just a step behind the cops and always in front of the camera, smooth-talking lawyer Pavel Astakhov became a household name through his work on the year's most politically charged cases. Vigorously arguing in the court of public opinion as well as the court of law, he was able to draw attention to the plight of his clients — and at the same time to himself.*

Shortly after Cheri's visit to me in prison, Pavel and Andrey Andrusenko came on board as my lawyers. As I would later learn, Cheri had met Pavel for the first time in the reception area of the Lefortovo lounge, in the company of Brad Johnson. Johnson had whispered to her, "Do you know who he is? Do you know his background? Do you know what you are doing?"

Cheri told Brad that she knew precisely who Pavel was and had agreed to have him represent me because the American embassy in Moscow, and the entire American government, had done virtually nothing so far to change the status of my case. Pavel's presence, Cheri hoped, would at the very least shake things up.

From the time of Cheri's visit, Pavel Astakhov kept in close contact with Congressman Peterson's office, both in regard to the

*Sarah Karush, "A Lawyer with Rhyme and Reason," *Moscow Times,* January 5, 2001.

interrogations and to his own behind-the-scenes efforts to obtain my release before any trial might begin.

A few days after Cheri returned to the U.S., Pavel, Andrey, and I met in Shelkov's office for the first time. The chief interrogator was all smiles as he lay in front of the lawyers the secrecy documents they had to sign in order to represent me. It was unclear to me whether Pavel and Shelkov had been aware of each other during the years Pavel had spent in the KGB, but Shelkov certainly knew Pavel by reputation.

Once Pavel and Andrey had signed the papers and left the room, Shelkov immediately dropped his affable demeanor and became enraged. Even my limited Russian was enough to permit me to discern the anger directed at Astakhov — and all lawyers who might interfere or intercede in the FSB's unholy mission of arresting and punishing whomever it believed to have broken any laws. Shelkov's bitter tirade filled me with contradictory emotions. On one hand, it gave me some hope that my new lawyers might have enough clout to win my freedom; but at the same time, I could not escape the sense that Shelkov and the FSB would relish a victory over Pavel Astakhov and would do everything possible to best him and that I might have to suffer for Pavel's effrontery.

One day without notice I was escorted down unfamiliar hallways and into what I recognized from prison chatter as "the bread truck," a van used to transport prisoners.* There was a cage inside the windowless confines of the van, and I was shackled in it for a short ride. I was never certain of my destination, since I did not see the exterior of the place where I was taken, but I believe it to have been Lubyanka, the KGB headquarters. There, in a large room, three women began to ask me questions in a perfunctory way, and I provided short answers.

*The idea was a holdover from the Soviet era: to transport prisoners in a vehicle that would blend in with the other traffic on the road and not identify itself as a prison van.

Most of the questions had to do with verifying the information about my career that earlier interrogators had collected; a few concerned my motivations. I surmised that this was a psychological examination and wondered if it was preparatory to having me committed to an asylum for the insane, a common Soviet-era destination. But after a few hours I was back in the bread truck, to return to my cell at Lefortovo as though the excursion had never happened.

Because the materials that had been in my possession did not yield very much that the FSB could use, Blubber-Butt was initially somewhat at a loss when interrogating me. That was partly the FSB's own fault: the initial investigators had been so unskilled that when they opened my Palm Pilot, they had taken the battery out, which erased the RAM files. But generally I carried nothing incriminating. My little video camera had mostly home videos, and my audio recorder, four blank cassettes. And as I had predicted, after the FSB made a closer examination of the stack of papers seized from me, they were put aside as less relevant than the interrogators had originally believed. I wrote in my notebook my distinct impression of the questioning thus far:

> Interrogators have given some subtle hints that they might welcome a pre-trial solution but the initiative must come to them from higher government levels. They could continue the current "investigation" activities for months/years and be happy; it represents credible work for a dozen or so of their people. They have no apparent pressure to conclude investigations in a timely manner, and my situation could not mean less to them. HELP!

But in July the dilatory questioning turned very specific as Blubber-Butt began to focus on materials that I had *not* carried with me to Russia but had turned up in the cache confiscated from Dan Kiely, chief among them the five papers on hydroreactive generators prepared between 1996 and 1999 by Anatoly Babkin.

I gathered from the gist of the questions that Kiely had been interrogated by the FSB and more or less forced by them to sign a confession that implicated me in espionage, in exchange for being sent home right away. I figured that the confession document was in Russian, which Dan could not read, and that he had probably been coerced into signing it and had very little notion of precisely what he was signing. But the "confession" was much less damaging than his sloppiness of having carried papers that I had pleaded with him to leave at home.

One can gauge the importance that the FSB placed on these papers from the later indictment and from the eventual verdict. I was never permitted to keep a copy of the indictment, but I do have a copy of the verdict; even in translation, its language is clear. I was charged with, and later convicted of,

> handing over to foreign organizations in the person of military-industrial establishment enterprises, collected technical information revealing the lines of research and development activities on M-5 "Shkval," an anti-submarine missile complex which is in operational service in the RF Navy.

Specifically, the charge was based on my obtaining the five Babkin reports. The first four were mailed to me (for transport to ARL) between October 1996 and July 1997, and the last was transmitted to me via e-mail on July 20, 1999. Typical of these reports was the paper on Component 2, Part 3, titled "Gas generators which use water as an oxidizer," which the verdict describes as dealing with "thermodynamic and stochiometric research results, the methods of calculating nozzles of missile engines using hydro-respondent fuel."

In my interrogation by Blubber-Butt — with the occasional participation of Shelkov, when the chief interrogator thought there was a particularly cunning trap to be sprung on me that day — I was asked in detail about all five reports. This was the toughest questioning since the ordeal had begun; what prepared me for it, I realized with a start, was not the training in resisting interrogations I had received prior to going to Sweden but the experience of fielding the sometimes harsh questions of three- and four-star admirals in the Briefing Theater of the Intelligence Plot. We briefers were taught to be confident, to be in charge of our information, and to withstand derision and doubts about the accuracy of the information we were conveying. Having successfully parried questions from tough admirals of the American Navy, I determined that any query put to me by that amateur questioner, Blubber-Butt, would not break me.

Blubber-Butt devoted an entire day's session to each Babkin report. Patiently, and for each of the five documents, I told Blubber-Butt and Shelkov (1) that we had obtained a letter from a Ministry of Defense official permitting us to work in the area of high-speed marine technology and had a signed agreement with Bauman to provide and vet the reports, (2) that all the reports had been described to us by their author and his university as containing only unclassified material, and (3) that the reports had been sent to us only after a security committee at Bauman had reviewed them to make sure that they did not contain any sensitive material. Babkin did nothing without the express permission of his superior, whose name and signature were on the agreement between Bauman University and Penn State ARL.

To buttress my assertions, I recalled for the interrogators an incident in November of 1999, when Dan Kiely and Anatoly Babkin had gotten into an argument. Twice during the discussion, Babkin had refused to answer a technical question because to do so, he said, he'd have to touch on classified information. When the objection came up a third time, I was satisfied with Babkin's refusal to answer on those grounds, but Kiely was not. Dan accused Anatoly of using the classification issue to avoid having to answer the technical question. The two of them had such a heated argument that it became necessary for me to ask Dan to stop his questions altogether and end the session, rather than have him press to obtain information that we were not supposed to learn.

Blubber-Butt dismissed my assertion that permission had been obtained from within the Ministry of Defense, contending that Region either did not cooperate with me or did not have permission to export materials dealing with the Shkval. He alleged that I had gone directly to Babkin and made a personal deal. Nonsense, I replied; Region had suggested we work with Bauman and had introduced us to the Bauman folks. Moreover, the payments for the first four reports had been made by wire transfer from Penn State ARL to Bauman, and the last one had been made by Bolshov, to Babkin, from funds I had sent Bolshov to pay various contractors.

There was evidence to support what I said, but the interrogators were uninterested in seeing it. Once more my lawyers offered the FSB a copy of the Ministry of Defense permission letter and the documents pertaining to permission from Bauman to commission the Babkin reports, but the FSB refused to accept them. About this time an article in a Russian paper reported that the acquittal rate in cases brought to trial on the basis of information obtained by the FSB was one out of two hundred, and after listening to the FSB refuse to look at exculpatory evidence that my defense offered them, I could understand such a ratio.

In April 1998 there had been a hitch in our ongoing agreement with Bauman and Babkin: the U.S. State Department rather suddenly imposed restrictions preventing any American government-funded institution from working directly with Russian institutes involved with supporting weapons development in Iran.* Bauman was on the blacklist. ARL could not permit me, even as an independent contractor, to deal directly with Bauman, so we had to find another way to continue the project. ARL and I worked out an arrangement whereby Babkin became a consultant to Bolshov's

*Peter Eisler, "Funding Ban for Russian Agencies a Delicate Issue," *USA Today*, April 16, 1998.

company, with the approval of Babkin's superior at Bauman, which I shortly obtained; that's how Babkin completed his last report. Blubber-Butt, of course, took this mild roundabout procedure as more evidence of nefarious activity.

Other materials seized by the FSB from Kiely's briefcase and the hard drive of his laptop computer were also used by Blubber-Butt to try to shake my story. These included technical reviews by ARL personnel of the Babkin papers, drawings of how the HRG generator would look as it was being tested — we had been trying to arrange for Bauman to test a simplified, scaled-down model of an engine on a stand — and papers analyzing the combustion characteristics of the hydroreactive fuel. Some of these papers, ones relating to the fuel, I had never seen before; Kiely must have obtained them without me. "Very damaging, perhaps classified," I worried about these materials in a note.

These were precisely the sort of papers that I had told Kiely not to bring into Russia because I feared the FSB would misconstrue them. They contained words and phrases that, although actually innocuous, could be pounced on by a security agent untrained in sophisticated technologies for *sounding* suspicious, reflective of high-tech robbery or espionage: underwater propulsion device; solid-fuel rockets; powdered-metal reactive engine; supercavitation.

And as I had feared, the FSB did misconstrue these words and the documents containing them.

Because the materials that the FSB considered most damaging had been seized from Kiely, not from me, they provoked an offhand comment from Blubber-Butt one day that the FSB had "kept the wrong guy," meaning that they should have detained Kiely for a longer term and sent me home right away. That hadn't happened, of course, and they weren't going to send me home now, because to do so would imply that they had made a mistake in their investigation.

Instead, Shelkov told me that all the seized documents were being submitted to a panel of scientific and security experts who would determine the extent to which the papers had damaged the Russian Federation. While that assessment was being undertaken, he said, there would be a hiatus in the interrogation process.

As happens in many cases of national and international importance, people began to expand their own agendas into the matter of Edmond Pope sitting in limbo. Those who wanted to denigrate the Clinton administration seized upon the lack of action by the State Department and the Clinton White House. One even went so far as to suggest suing the State Department and Russia for collusion in failing to maintain my health in prison. Others sought to use the Navy's apparent inaction to reinforce past charges about what they believed was the Navy's bad behavior. Some individuals wanted to make themselves into heroes by bringing me home, or to have their organizations become instrumental in doing so, when they had neither the connections nor the clout to do more than muddy the waters. Finally, a few reporters saw the "story" as a route to aggrandizement.

For the most part, though, a great number of people tried legitimately to help, only to find that getting me out of Lefortovo was an inordinately difficult task. And this, for a group that included heavy hitters such as the current and two former Directors of Naval Research; four former Directors of Naval Intelligence, two of whom had also been assistant directors of the CIA; the man who had been in charge of the Situation Room in the Reagan and Bush White Houses (former naval Captain Dave McMunn); and guys like Ted Daywalt and Dave Moss, who used their extensive networks of contacts in news organizations and on Capitol Hill.

There were the inevitable crossed wires. Bill Daniels at State, who had initially been unsympathetic toward Cheri, was inclined to give Keith better, more detailed, and somewhat sympathetic reports from Moscow. People at ONR were able to obtain from State the written reports that consular visitors to me in jail were supposed to make, before these reports were given to my family; in some instances, ONR was able to suggest to its former leader, my partner Brad Mooney, that he advise Cheri to ask for the reports by date and number. The several congressional offices were not always in sync, though all were pointed in the same direction.

When Cheri returned from Russia, she went immediately to the funeral of her mother. On her way to San Diego, she sent messages to our family and friends: unleash the Southern strategy. By this she meant asking all supporters to do what they had previously and purposely refrained from doing — call and write every representative and senator they knew, asking them to pressure the State Department, the White House, and the Department of Defense to get me out — and also to vote for Congressman Peterson's nonbinding "Sense of the House" resolution, which would tell Putin that, as the congressman insisted, "It is time to send Edmond Pope home." The resolution stated that no further American funds ought to be provided to help the Russian Federation join the World Trade Organization; that no funds be provided to Russia through any arms of the American government; and that the president and Congress should "oppose further loans to the Government of the Russian Federation by any international funding institution of which the United States is a member."

Some of Peterson's constituents and colleagues who generally supported him chided him for sticking his neck out for me. It was, they insisted, a politically risky thing to do, because if I actually turned out to be a spy, Peterson would lose stature at home and in Congress. He later recalled his answer: "I'd tell them that my best information was that Ed Pope was not a spy, but even if he was, I'd try to bring him home. I'd ask them, 'What if it were your son, or brother, or father? Would you want me to work on his behalf?' "

Peterson's bill, Congressional Resolution 364, would henceforth be the rallying point for support in the United States for my release. Several dozen other members of the House quickly signed on as cosponsors. Eventually, the number of cosponsors would reach 140, through the efforts of friends and family members such as my son Brett in South Carolina; my sister, Brenda, in Denver; my mother in Oregon; distant cousins in Oklahoma, Texas, Alaska, California, Washington, and Missouri; school classmates from Grants Pass; people I'd met on my tours of duty; neighbors; and friends of friends, all of whom used their connections, knocked on doors, and spent many hours on the phone to urge their representatives and senators to back John Peterson's resolution. A website to monitor its progress was built and put on the Internet. Such lateral groups as the Retired Officers Association and the National Military Intelligence Association, brought into the picture by NIP directors, began to feature my plight in weekly newsletters, and it was discussed at a national conference of the Veterans of Foreign Wars. Eventually, the list of organizations that sent out bulletins about me to their members included the Navy League of the United States, the American Legion, the Order of the Purple Heart, the Naval Reserve Association, the Naval Enlisted Reserve Association, and the Reserve Officers Association.

The second strand of the strategy was to aggressively court public opinion through the news media. This would put additional pressure on the State Department and the White House. Newspapers that called the State Department about me in late June were still receiving the response, "We are not actively seeking Ed Pope's release," which infuriated my family and supporters. They hoped that if State Department spokesmen had to answer reporters' questions about me in daily briefings, if the switchboard at the White House became swamped with calls asking that the president work harder to have me released, if congressional offices were flooded with mail and e-mail demanding that the representative or senator vote for 364, then things would happen to accelerate my release.

It was also a matter of generating repeated and overlapping messages. When a high-ranking official in the State Department or the White House received one call from a well-placed person asking that his or her agency do more for Ed Pope, he or she could ignore that request or pass it off to a junior staffer for a formulaic response. But when similar calls came from several people, each in a different walk of life — one from the Navy, another from a lobbyist, a third from a substantial donor to a political party — and all of the callers were influential people whom the official trusted, the official would begin to understand that the message they conveyed was too important to too many people who counted, and therefore it could not be indefinitely ignored or foisted off to an underling.

As Cheri had promised, at two in the afternoon, no matter where she was or what she was doing, she would take ten minutes out of her day to commune with me at long distance, even leaving meetings to do so. But then, an increasing fraction of her days were becoming consumed with projects to assist me. Among the more onerous and time wasting of tasks was fulfilling the Russians' demands for official copies of important documents. Notarization was not enough for the FSB; they insisted that certification be supplied by state officials. So Cheri spent a day or more per week traveling to Harrisburg, capital of Pennsylvania, and having things properly signed and sealed. Often, though, the Russians would find some small thing wrong with the document and refuse to accept it, so that she would have to redo my medical records, my father's medical records, papers about power of attorney. No detail was too small for the Russians to overlook in their zeal to obstruct the attempts of my family and friends to help me.

Coincidentally, the president was scheduled to be at State College for the July 12 meeting of the National Governors Association (NGA) that Cheri had been helping to organize. During his governorship of Arkansas, Clinton had been a leader of the NGA and had used his activities with the group, such as championing education reform, as a platform for his 1992 campaign. The 2000 NGA meeting would be the last to be held while Clinton was president. But until early July, the White House resisted suggestions that the president meet with Cheri in her hometown.

Cheri had been working for two years to coordinate this NGA meeting, and some of her work had involved correspondence and phone calls with the White House and the Secret Service. At the same time, of course, she had been peppering the White House with missives asking Clinton to push Putin for my release — for instance, by getting the Pope case on the agenda for the second Putin-Clinton summit to be held in Okinawa during an economics meeting toward the end of July. The clash of Cheri's two agendas evidently alarmed the Secret Service enough so that a week before the NGA meeting, they began to keep her in their sights at all times. "I think they were afraid I might do something wild," Cheri recalls.

Nudged by Keith McClellan, the *Centre Daily Times* rhetorically asked whether the president would meet Cheri during his visit to the campus. Then reporters for other news media began to put the same question to the White House schedulers. Within a day, the White House staff finally suggested a meeting; the staffers first proposed to Cheri that she sit at the dais when Clinton was speaking, so that he could acknowledge her in his speech and thereby give recognition to my cause. She refused that honor. Next, Clinton's schedulers offered her a photo opportunity with the president. She refused that honor, too, and told them there was probably no point in her meeting the president since the White House had already said it was doing all it could — and there had been no progress toward my release, anyway. Finally, the schedulers almost begged her to accept a short meeting with the president. Cheri agreed on condition that the meeting be private, with no photographs taken, and that there be at least one other person in the room. The schedulers suggested Mark Medish, the National Security Council staffer with whom Cheri had had some contact; she didn't like the idea, because she believed (whether rightly or wrongly) that Medish had manipulated her to stay quiet prior to the first Clinton-Putin summit, but she acquiesced.

The president was spending most of that week at Camp David, working for a breakthrough in the Middle East negotiations with Prime Minister Barak of Israel and Palestine Liberation Organization chairman Yassir Arafat. He had very limited time at the Penn State NGA meeting. Nevertheless, a face-to-face meeting between my wife and President Clinton was scheduled for a small and undistinguished antechamber to the banquet hall in the Penn Stater Conference Center Hotel. Medish and Cheri had to wait there for a while until the president arrived, and they were glaring at one another in silence when the president finally walked in.

President Clinton did most of the talking, Cheri recalls. She didn't like him, but had to admit that he was attractive and magnetic. He told her that some slow progress was being made on getting me out of prison in Russia and asked how she and her lawyers thought he ought to proceed from here on. Should the White House make more of a public outcry and embarrass Putin or continue behind-the-scenes to pressure Putin for my release?

Taken aback, Cheri said in response, "You're the leader of the free world, I'm a nobody, and you're asking *me* for advice?"

The conversation went around and around on this subject for a minute or two. Cheri wondered to herself whether the president might be pressuring her to make a decision on the strategy so that if it later went wrong, he could avoid the blame. But she did not waver: the decision on how best to proceed to win my release, she insisted, was the president's, not hers. He again offered the opportunity to have her picture taken with him, and she refused. In closing, Cheri said, "I'm not going to give up until I get him back. No matter how difficult. I'm not going away."

"You're one strong lady," the president responded. He concluded the meeting by saying that he would raise my unjust incarceration in Russia with Putin at the Okinawa summit, and he believed this would result in my freedom. "Putin owes me one," the president told Cheri, but did not further explain this cryptic remark.

Cheri had perhaps hoped to keep her meeting with the president quiet and private, but Clinton's press officer came out and told reporters what had been said, and put some spin on it. The next day, the *Centre Daily Times* printed an editorial: "Congress must butt out of the Pope affair." The nub of the paper's argument was, "This case will be resolved more quickly without Congress trying to decide what course U.S. foreign policy will take on the basis of a single dispute that can be settled without the fanfare of a showdown involving Congress and the Russian government." In the wake of Clinton's meeting with Cheri, this was precisely what the president wanted to have the public think.

According to well-placed sources on both the American and Russian sides, by the July 12 meeting between the president and Cheri, the administration believed it had a deal in place to obtain my release in time for me to return home to attend the August 5 wedding of my son Dustin. President Putin's men inside the FSB had reported to him that the evidence against Ed Pope was nonexistent. Others in the Kremlin had successfully argued that the government had already wrung out of the situation any positive public-relations value that it might obtain and that from here on in they would lose rather than gain public acclaim by keeping Ed Pope incarcerated.

The Okinawa economic summit of the G-8 was the next scheduled meeting between Putin and Clinton, and Pavel told me in a whisper at Lefortovo that the deal was to be consummated then.

Having this possibility of release in the immediate future, I was in agony as I watched Russian television, night after night, showing Putin making his way to that summit by way of India, China, and North Korea; on those stops, he signed arms sales deals and in India obtained the release of five Russian prisoners. I hoped the subject of the next report about prisoners being released would be me.

Putin's summit with Clinton at Okinawa was very brief. The American president had delayed his departure for the Japanese island until the last possible moment so he could spend his time at Camp David with Barak and Arafat. At the G-8 meeting, what with their other official duties, Putin and Clinton spent only 45 minutes together in a one-on-one session. Communiqués later said that the two leaders discussed the antiballistic missile defense shield that the United States wanted to begin, and nuclear arms reduction. Clinton was reported as having scolded Putin about constraints on press freedom in the Russian Federation. No public mention was made of any discussion between the presidents about my case.

This did not exactly accord with the plan that Pavel had vouchsafed to me in advance of the summit: that Clinton would cement the deal with Putin in Okinawa, then hold a press conference to discuss my case and the issue of freedom of the press in Russia.

When there was no press conference, I thought they were just being crafty, keeping mum for a while until the release could be arranged. But then nothing happened, and I was crushed.

According to other sources with whom I later spoke, the leaders did discuss my release at Okinawa. That is all the hard evidence I have, but my guess is that there was no press conference after their meeting because the proposed deal was amended: Putin agreed to go home and release me if Clinton did not embarrass him publicly by holding such a press conference. In any event, according to several sources, when President Putin told the FSB that he was going to release me, the upper hierarchy of the security agency had a fit. FSB director Patrusev adamantly insisted to Putin, in a private meeting, that Ed Pope was a spy and that the FSB could prove it in court. Moreover, Patrusev said, if Pope were to be released now, the FSB would appear foolish, and the action would undercut the FSB's demands for higher appropriations, etc. Putin decided to acquiesce to the FSB's insistence to put me on trial, but said that he would release me after a trial and conviction. Putin reportedly told Patrusev that if the FSB did not properly manage the show trial, he would strip Patrusev of his high rank, bust him to a uniformed man, and have him sent to the front lines in Chechnya.*

*It's a great story, but I don't believe it's the whole tale; more likely, Putin was more involved than the story implies; he and Patrusev probably collaborated without rancor on a plan to drag the case through the courts in a leisurely way, but to make it appear that Putin was displeased with the amount of time the trial was taking.

* * *

I had begun to complain to my notebook of headaches so severe that I could not think straight and of suspicious-looking new growths on my neck and face that the FSB doctors refused to take seriously. On July 20 — about the time of the Okinawa summit — my medical condition took a sharp turn for the worse. I had been imprisoned for more than 100 days, and all previous episodes of discomfort had always passed. This day I began to suffer from stomach and digestive cramps, headaches, and nausea that would not fade. When I told the guards I was too ill for an interrogation session, they forced me to go to Blubber-Butt's room anyway.

Blubber-Butt tried to present me with "evidence" I hadn't seen before — more from Kiely's stash — and when I asked lawyer Andrey what to do, Blubber-Butt launched into a tirade, forbidding us from talking and Andrey from advising me; the interrogator got so steamed that he left the room. Later, things calmed down, though Blubber-Butt still would not let Andrey advise me on responding to the new evidence.* Andrey said it was my right, however, to protest that I was too ill to answer questions. This I did, and despite Blubber-Butt's annoyance, he had to send me back to my cell.

When the Okinawa summit passed and there was no movement toward my release, my mental and physical health declined. In the last ten days of July, I became increasingly ill. The nausea, cramps, and blinding headaches intensified. The incessant smoking by the other men in my cell grew more and more noxious, irritating my nose and eyes, permeating my clothing. I asked to have a prison doctor visit me in the cell, but all through the weekend no one came. Pains in my kidneys added to my other woes. The following Thursday, on the way back to the cell from the steam baths, I almost col-

*Shelkov later told Brad Johnson that while suspects under interrogation must have an attorney present during their questioning, they could not consult that attorney before answering.

lapsed. My vision blurred; I felt feverish. The prison doctor — who finally saw me — told me that my vital signs were normal. He advised that all these symptoms were due to a "nervous breakdown," gave me an over-the-counter analgesic and some vitamins, and suggested I consume dairy products. When Brad Johnson next visited me, I could not recall for him the titles of the more than thirty books I'd read — earlier, I'd been able to rattle them off. I told him I had not the attention span to read or to watch television, nor the physical stamina to exercise.

July 25 was "the most difficult single day of my captivity," I recorded in my notebook. Unable to eat during the previous two days, I was ravenous, but when I tried to eat my stomach swelled and ached. The lack of action emerging from the Okinawa summit and my isolation from news of efforts on my behalf were agonizing. Then five letters from home were delivered to me, and their combined emotive power hit me like a thunderclap. "I Must Be Strong!!!" I wrote in my notebook, but I felt very weak.

On July 27, I was so sick that they took me to the doctor, but nonetheless also brought me to the interrogation room. Blubber-Butt enjoyed my distress and told me, "August will be very hot, entire month forty degrees," referring, I was certain, to the intensity of the interrogation more than to the physical temperature.

When Brad Johnson visited me in prison again, my health had so visibly fallen off that he wrote to Washington with alarm: "Pope was noticeably thinner than he appeared at the last consular visit and throughout the meeting exhibited slow speech and thought." He phoned Cheri directly to say that I was extremely ill and that he feared for my life.

On July 31, another American citizen died in a Russian prison, though not in Lefortovo. This guy was only fifty-one. The cause of his death was complications of diabetes, brought on by the refusal of the Russians to allow his medication to be given to him. I did not learn of this incident for some time, but the American embassy found out immediately after the death; it contributed to their alarm over the state of my health and of course to the fears of Cheri and my family and friends in the United States.

Each day, toward the end of July 2000, I felt physically worse. I began a medical journal, the handwriting disjointed, the notes filled with unexplainable symptoms. There were two full days during which I could neither eat nor stand up. Was I gravely ill? Was I being drugged? I fell into a severe depression, alternating between fitful sleep and pain-filled wakefulness, descending further and further with each hour. No matter what I said to the interrogators, no matter how hard my partisans at home tried to spring me, my status was unchanged; I was going to be tried, convicted, sent to a real prison for twenty years, and there seemed nothing that I could do to affect that course of events. My thoughts went in circles or in ruts that I could not escape. I considered suicide. I spiraled lower and lower into the mental chasm, uncertain as to whether I would ever be able to climb back to health and steadily losing hope that I would ever return home — alive.

CHAPTER 7

August Is Very Hot

As a Naval Intelligence officer, I was trained to analyze complex and unclear situations by banishing emotional judgments from my mind and trying to coolly assess the quotient of truth and threat in each piece of information.

I have tried to apply that technique to the source of my deep mental and physical illness in Lefortovo that began in late July 2000. American doctors to whom I have spoken generally dismiss Graves' disease, which often brings about severe weight loss, as insufficient to explain the multiple problems that became so severe in a very short period of time. Other possible medical explanations include a virus, perhaps a hepatitis from infected food. Such a cause is probably ruled out by the fact that I later, and rather quickly, recovered from the most serious aspects of the illness. A third explanation: drugs in combination, put in my food by cell mates or guards working for the FSB — or for some other entity.

Four snippets of information support the last theory: (1) A few days prior to the rather sudden onset of my illness, the assistant to the warden of the prison, a lieutenant colonel, came into our cell and asked the older man in the cell, in English, if he was Mr. Pope. Startled, I identified myself, and he told me what his position was, expressed the hope that I was being treated well, and said that if I encountered any problems I should send him a note and that permission would be granted for me to come and see him. I thought this strange at the time.

(2) Reports coming out of Lefortovo to American authorities in late July said that I was ill but was not in danger of dying. How would anyone be able to make such a diagnosis if they did not know the source of my illness? (3) The arms dealers/contractors involved with the purchase of a Shkval had conversations in July with Peterson's office, in which they suggested that I could be sprung from prison by the proper payment of the commission to the FSB on their aborted sale; they also claimed to be working on getting me out by other means. (4) Early in August, someone from the office of the Secretary of Defense called to inform Peterson — the congressman had not heard from the Pentagon directly since the beginning of my incarceration — that attempts to get me out through their contacts had failed and that he should redouble his efforts.

These strands were woven together when, after I returned home to the United States, a variety of sources told me that in July, in addition to the souring of the deal between Clinton and Putin to release me in time for Dusty's wedding, there had been a failed plot to obtain my release by force. The idea was to make me so sick that I would have to be taken to an outside hospital; on the way there, the bread truck would be ambushed, and I would be taken directly into the American embassy or to some other safe location, and from there would have been spirited home.

If there ever was such a plot — and the evidence suggests that there was (and that someone in the prison was involved) — it would surely have ended in disaster. If I were to have been killed in the ambush, or if the ambush succeeded and I had been taken to safety, the attempted rescue would have proven my guilt in the eyes of the Russians. Fortunately, this fiasco-in-waiting never took place.*

Who could have slipped me the drugs? By a process of elimination, I've concluded that if anyone did so, it would have had to have

*Had the FSB or Putin merely been looking for an excuse to release or transfer me to a hospital, there would have been no need to literally make me sick. Insuring that government statements were backed up by facts was never a priority in Russia.

been Sasha, a cell mate who had become my friend. Three small pieces of information support the notion that he was working against my interests: (1) He would not give me his address, while all my other cell mates willingly gave me theirs, and only after much discussion did he give me his phone number, writing half of it in one place in a book I'd been sent and half of it in another place in the book — a neat piece of spycraft. (2) We planned to meet in Italy when we got out, at which time, he said, he'd be able to tell me lots of things that he was unable to tell me while in prison. (3) After I had begun to recover, I was moved to another cell, but Sasha was not, and thereafter I never saw him again during my incarceration.

I asked another cell mate for the Russian phrase for "I do not feel well" and wrote out the transliteration in my notebook: *Ya ploho sebya chustvuyu.* I was certainly sick enough to alarm everyone in Moscow concerned with me — those few people on the staff of the American embassy, an institution that prior to this time had been notably complacent about my condition.

Brad Johnson's report of his August 1 visit to me, the sense that the Clinton-Putin deal for my release had unraveled, and pressure on State by the media generated an unusually forceful meeting in Washington, D.C. on August 4. Assistant Secretary of State Mary A. Ryan called in her opposite number, Russian chargé d'affaires Igor Nemerov, to protest the death of the other American in a Russian prison on July 31, to express grave concerns about the state of my health, and to demand that I be released before I, too, became a casualty of the Russian prison system. Ryan also delivered a formal note of protest. That note insisted that the Russians permit consular officials to bring in an American doctor to determine more accurately the state of my health and protested the "arbitrary denial of food deliveries, curtailed mail, and access to attorney during interrogation." State Department minutes of this meeting reported that Ryan "stressed [that my] continued incarceration represented an unnecessary irritant in our bilateral relationship." She asked if Putin had "reviewed" the matter, as he had told Clinton he would. She stated that I had violated no law and that I was sick and should be released. She warned Nemerov that the Pope case had to be resolved before the next meeting of Putin and Clinton.

The Russian Ministry of Foreign Affairs's formal response to the American diplomatic note was more sand in the eyes: the Russians would not permit an American doctor to see me in prison and certainly not to draw blood, as had been requested by the United States.* They also expressed concern that Cheri was to appear on ABC's *Good Morning America* and on other national television programs, which in the Russians' view would reflect unfavorably on both the Russian and the American diplomatic efforts to resolve the situation.

Cheri's appearance on *Good Morning America* was part of her response to the call that had come from Brad Johnson in Moscow after his visit to me on August 1, in which he told her he feared for my life. That call, and a second one from the State Department in Washington echoing Brad's concern, was a turning point for Cheri. After it, she determined to pull out all the stops — to do whatever she could in order to exert more pressure on the parties to get me out of Russia. Our sons recall her at that time as so fearful for my health, and so intent on helping me, that she was unable to give her best attention to Dusty's wedding on August 5.

Her determination included agreeing to make television appearances that she had previously not wanted to endure. When she appeared on *Good Morning America* with our elder son, Brett, the interviewer seemed appalled by her stories, such as that of having had to initially call the State Department for news of my arrest. Cheri followed this with other TV sessions. These programs, along with articles in the Reuters and Associated Press news wires, continued the

*Rear Admiral Sumner Shapiro, another former DNI, recalled that he and several shipmates had been drugged by the Soviets in Odessa in the 1960s, and the United States had been able to prove that by analysis of the naval officers' blood and urine samples. That capability, the admiral believed, was the reason behind the Russians' refusal in my case to allow an American doctor to take my blood while I was in prison.

pressure. ABC and Peterson's office reported floods of calls from people who had seen the reports or read about them in newspapers and wanted to help. Some stations showed, on-air, the phone numbers for the White House and for Secretary of State Albright, which also spurred many calls to those government entities. Some days after this burst of publicity and pressure, the Russians relented a bit. Shelkov sent me a note saying that an American doctor and my lawyers would be permitted to stand and watch as the Lefortovo doctors performed a physical exam. That never happened. But the Russians did let an American doctor see my prison medical records and make suggestions to the Russian prison doctors of new tests that they could do on me that could reveal the extent and source of my illnesses.

But they didn't give in on the larger issue: my continued incarceration. My imprisonment and forthcoming trial still served the interests of Putin, his chief advisers, and the FSB, which in early August was just winding up the interrogation and investigation phase and was about to prepare for my trial. To halt this justice-system process at that juncture, the FSB must have convinced Putin, would embarrass them and undercut the public-relations value of pursuing the "American spy" through the courts.*

Also, my release just then would have focused attention on the Shkval torpedo that I had been accused of stealing — and this was just the moment when the Russians were finalizing arms sales deals to Iran and to China, the latter including sales of Shkval torpedoes. The Russians wanted the hard cash that these deals would bring, but did not want the deals to be examined too closely or highlighted by the West, as they certainly would have been had I been released close to the time the sales were consummated.

Though I remained ill and almost unable to follow the interrogators' questions, I stopped considering suicide after Blubber-Butt

*In support of a hard line on justice, on August 14 Putin appointed career KGB man Yuri Demin as first deputy minister of justice; Demin had been chief of the FSB legal service and chief military prosecutor.

Petukhov demanded that I confess. He had made the same request several times in the past. This time, something gave me the strength not only to refuse but to laugh at his demand, which visibly upset him. It was apparent to me that his blustering insistence on a confession had worked for him in many other cases, and he could not understand why I did not crumble under his relentless attack. My laughing refusal, and the continuing need to keep fighting this bastard, shook me out of my depression. And even in my weakened condition I was aware that Blubber-Butt was rushing now, glossing over the questioning in regard to several documents that were equally as pertinent as those that had been shoved under my nose in prior sessions. That must mean we were nearing the end and were working toward a deadline that was political and not connected to the presence or absence of real evidence in the case.

I was correct: in the second week of August I learned from my interrogators that this phase of the investigation was winding up and would conclude with the findings of a report from the special commission of experts established to examine the five Babkin/Bauman papers. When it finally arrived, on August 10, I was told that I would be allowed to make a complete and accurate English written translation of the document and to take it to my cell to work on. But the next morning, Shelkov told me that the document produced by the commission was classified, and my lawyers and I could only look at it in the interrogation room. The next day, we began to look at the commission's report.

The first shock came from the name of the leader of the commission, G. V. Logvinovich. I knew him, and he knew precisely what work I had been doing, because he'd been part of it. He was an expert on hydrodynamics whom I had first contacted in 1996, when someone at ARL had mentioned him to me and suggested that I ask him to be involved. We met in July of 1999, at a sit-down with Russian Technologies and Region in which the parties agreed to bring to market a broad range of technologies developed in high-speed underwater research; Logvinovich's name was also on the proposals passed back and forth in January of 2000, since he was one of the investigators who would do the testing. Now here he was, chairing a commission to justify the conclusion that the Babkin reports and other papers seized from Kiely but attributed to me dealt with classified material! In condemning me, he would in part have to condemn himself — a tactic reminiscent of the Stalin-era purges and show trials. Either Logvinovich had been forced to sign a report written by others or had been coerced to write it himself. I immediately informed Shelkov that I knew Logvinovich and that he had earlier agreed to work for me. I had even had in my wallet the business card he had given me, with his name in English. "Yes, we know that," Shelkov said, "but we trust him."

A second surprise came from an item glancingly included in the report, one that I had not been aware of and that I believed the U.S. intelligence services were also unaware of: that Russia had sold fifty Shkvals to China. I had earlier been told that this information was classified in the Russian system — and here was the FSB, casually revealing it to me.

My lawyers and I were just in the process of going through the commission's report when the *Kursk* took over the headlines. On August 12, the Russian submarine *Kursk* sank in 350 feet of water in the Barents Sea, with 118 people aboard. The event was a Russian national tragedy and an enormous embarrassment. An Oscar II class type 949A nuclear-powered submarine generally armed with a variety of cruise missiles and torpedoes, the *Kursk* had been commissioned in 1995. Its homeport was Murmansk, on the Kola peninsula at the southern rim of the ice-free Barents Sea. The peninsula has several submarine bases and shipyards, along with at least 100 decommissioned nuclear subs in the harbors and, on land, 50,000 spent nuclear fuel canisters from reactors. The area is remote from Moscow and, reports suggest, had been a dumping ground even in the Soviet era.

I knew some things about Russian submarines that most people in Rússia didn't, such as that in the 1990s, Russia's aging nuclear submarine fleet had dwindled to the point where only fifty were considered operational, and most of these were idled for repairs, crew shortages, lack of spare parts, or operating funds. Some were not being operated in order to preserve their capabilities for times when they were urgently required. Oscar II class submarines are among the world's largest, 560 feet long and 78 feet wide, weighing 14,000 tons and constructed with double hulls and five separate pressurized vessels within the hulls. Each Oscar II submarine is so well constructed that it is considered able to withstand even the direct hit of a single torpedo.

It was later learned that in the days prior to the sinking, the *Kursk* had been taking part in naval exercises and was observed that morning by U.S. vessels to fire a cruise missile and hit a target 200 miles distant. She was known to be carrying two dozen cruise missiles and twenty-eight torpedoes.

Nothing of what happened was publicly known on August 12. The Russians only admitted on the 14th that the sub had had an accident on the 12th and was missing. Later, more detailed reports in the West concluded (from sighting the extended periscope on the sub, easily visible in pictures from underwater cameras) that the *Kursk* had been at periscope depth — the depth from which torpedoes are usually fired — when the accident occurred. At 7:28:27 GMT, Norwegian and U.S. Navy sonar operators had picked up a powerful blast coming from the area of the sub. At 7:30:42, there had been a second blast, so large that it almost deafened listeners at the sonar equipment. The tape was analyzed and revealed that the second blast was actually several nearly simultaneous explosions, equivalent to several tons of TNT, and of a magnitude great enough to register 3.47 on the Richter scale on seismic recorders in Norway, two thousand miles from the submarine. Norwegian seismologists later said this was the largest explosion they had ever recorded.

Since the Russians claimed at the time of the event, and for a long period thereafter, that the *Kursk* sank because of a collision with another vessel, it is important to note that no collision could have damaged the *Kursk* enough to immediately blow out its first four compartments, and no collision could cause a blast large enough to nearly deafen sonar operators and register on recorders 2,000 miles away.

Approximately one-third of the men inside the *Kursk* survived the initial blasts. As scribbled letters found in the pockets of the dead attest, some of the men lived on for two days in the darkened submarine, which fell 350 feet to the ocean bottom, until their oxygen ran out. During the period when it was not known whether there were any survivors, though the probability existed, the Russian navy refused offers of assistance in submarine rescue from the United States, Great Britain, Norway, and other countries.

My cell mates were scandalized that in the immediate aftermath of the accident, President Putin remained on vacation in southern Russia and did not rush back to take charge of any attempted rescue mission. Even those who knew little about military matters could tell that the government's statements about what had happened didn't add up. Quite quickly it became apparent to my cell mates, as it must have to most Russians who watched the news reports on television, that the Russian government was lying about what had happened to the *Kursk* and what could be done to extract any crew members who might still be alive.

One evening, the Russian news broadcast had file footage of the torpedo bay of a submarine, and I realized with a start that at least one of the torpedoes shown being loaded was quite different from the others and that it was likely to be a Shkval. I suppressed a bitter laugh: this was the first time I had ever actually seen a Shkval; before this moment, all I'd had to go on was a photo in a brochure.

As soon as I heard of the sinking of the *Kursk,* I realized it was bad news for me. The crimes of which I was accused had to do with torpedoes, and since I actually knew a good deal about torpedoes, I thought it highly likely that the *Kursk* "accident" had been caused by the misfiring of a torpedo — and concluded that if it had been, the authorities would want to come down all the harder on a foreigner accused of stealing torpedo secrets. I might well become a scapegoat for the sinking of the *Kursk*. This feeling was reinforced when, shortly after the sinking, the director of the Central Intelligence Agency, George Tenet, arrived in Moscow; he and his press secretary were at pains to point out to the Russian media that this visit had been arranged some time prior to the big news event. Nevertheless, the Russian press and population saw something sinister in the timing of the CIA director's stay in Moscow.*

Military experts believed that the *Kursk* was carrying both upgraded Shkvals and the more secret Stallion missiles, all of which are rocket-propelled. A former vice president of the Russian Federation who became governor of the Murmansk region later told reporters that two high-ranking military officers had informed him that on board the *Kursk* were civilian military hardware experts, sent along to test new torpedoes. Other, later reports also suggested that a Chinese observer had been aboard. It was confirmed to the press that the Russian civilians were the chief engineer and weapons designer at the Dagdizel military plant. "These men were supervising a regular test launch of torpedoes on the *Kursk*," the head of the plant later told the London *Times*. And the Russian defense minister confirmed to that newspaper that "the submarine's objective was to launch a cruise missile, and then . . . to . . . hit the main target with a torpedo salvo." Even the former submarine commander and regional legislator who chaired the official Russian committee investigating the accident later stated that the *Kursk* had been firing a torpedo or missile when it purportedly collided with another ship.

*Tenet was reportedly in Moscow to discuss matters of mutual concern with his opposite numbers in the Russian secret service; it is even possible that he was going to tell them formally that I was not employed as a spy by any American espionage agency — but because of the *Kursk*, he probably did not mention me at all.

A rocket-propelled torpedo is initially expelled from a tube in the hull of a submarine as though it were an artillery shell; then, once it is in the water, the rocket-propulsion system kicks in and thrusts it toward a target. According to an article by a submarine specialist in the Russian military newspaper *Red Star,** in 1998 the *Kursk* was refitted with a new and potentially dangerous torpedo-launching technology. This was done over the objections of the Navy and supposedly because the new launching system was cheaper than the old one. The trigger mechanism, when pulled, causes liquid fuel to ignite, producing a gas stream that propels the torpedo out of the tube. That liquid fuel was also reconfigured, substituting one that has a low flash point and can be very unstable unless certain safeguards are used. That was the reason that Western countries' navies had considered but then rejected the use of these fuels for the expulsion of their own torpedoes.

My guess (and that of other experts) is that in the process of test-firing one of these torpedoes aboard the *Kursk*, the trigger mechanism was pulled but for some reason did not propel the torpedo out of the narrow tube; with the torpedo stuck in the tube and within the hull, the time-delay mechanism then kicked in and fired the on-board rocket. Such a sequence of events would certainly have produced the first explosion, involving the unstable launching fuel, and the second set of near-simultaneous blasts, when the rocket ignited and blew up that torpedo and several others nearby. The series of explosions would have fatally damaged any submarine.† Examination by divers

*The article by Vladimir Gundarov appeared on the website of the *Red Star* on August 18, 2000, but only in an early edition. By evening, according to visitors to the site, the article was replaced by another that made no mention of the change in torpedo-launching technology.

†In February 2001, Vice Admiral Valery Doroguine, a member of the Russian inquiry commission, confirmed that two torpedo explosions had sunk the *Kursk*. But he said that the reason for the first explosion remained unclear. In May 2001, the German television channel ZDF cited a source in Russia as having said that prior to the last voyage of the *Kursk*, a dockyard procedure had failed to lift a defective torpedo from the submarine, and speculation centered on that particular torpedo as the source of the accident. (Report by Agence France-Presse, May 9, 2001.)

did reveal that the bow torpedo section of the *Kursk* had indeed been blown open from within.

During the news coverage of the *Kursk* disaster, one of the prominent interviewees on Russian television was the designer of the Oscar II class subs; I had met him, under difficult circumstances, seven years earlier. It was during one of my first visits to Russia, in 1993. A Russian submarine named *Komsomolets* had sunk in the Norwegian Sea in 1989. It was an experimental, one-of-a-kind nuclear-powered sub, and it had gone down in an area where the bottom was 5,000 feet beneath the surface. That was too far down for the sub to be rescued or raised, and as a result, the Russian government appealed to the rest of the world for assistance in attempting to encapsulate the *Komsomolets* so that there would be no leakage from its nuclear components; the Russians wanted other countries to contribute substantial sums of money for this purpose.

A senior admiral in the U.S. Navy, learning that I was on my way to Russia, asked me to carry a personal and very tough message from our Navy to the designer, and arranged for me to meet with him. I did so. The message I carried was this: because the Russian navy continues to build new nuclear submarines instead of using its funds to clean up the mess made by its older and decommissioned subs, the United States Navy will not contribute money to encapsulate the *Komsomolets*. The designer heard me out, but very quickly became enraged at the notion of the U.S. Navy telling the Russian navy to set its house in order. We did not part as friends.

Three days after the *Kursk* sank, on August 15, Shelkov told Andrey and me that the investigation phase of the inquiry was now complete and that there would be a trial in a month or so. Ten massive volumes of "evidence" had been produced, and we would be permitted to go over them to prepare our side of the case. He wanted me to sign a document saying that the protocols were complete and the court case could now go forward. I refused to sign it on three grounds: that my senior lawyer, Astakhov, was not there; that we were not permitted to comment on the protocols and include our objections to them — a promise they had made to me many times; and that the FSB had not prepared and presented a medical evaluation protocol, even though they had appointed a special medical examiner weeks ago, supposedly to complete such a protocol. Shelkov became more and more angry at my refusal to sign and eventually stomped out of the room.

That evening, he retaliated. I was moved out of the two-man "penthouse" in which I had been held during the latter part of my illness and back into the six-man cell. But I already knew some of the other men in that cell and immediately discovered that one of them had received permission to keep a refrigerator, which was a godsend in helping us to store fresh produce. The small victory I had won over Shelkov, and the presence of that refrigerator, spurred me toward recovering from my illness.

CHAPTER 8

Life As a Spider

The intimidating bulk of the ten volumes of evidence, I presumed, was intended in part to make me realize that such a weight of evidence would inevitably result in my conviction. It did not do that. I had resolved to fight, not to surrender and start to die; and the evidence books were a tangible adversary. Counterpunching was difficult, though, because my lawyers and I still did not know the precise basis of the charges. A few days after the volumes were given to us — and four months after I had been first detained — my indictment was handed down; we still were not permitted to read it, though, so we had to search through the 3,000 pages to figure out the bases for the indictment and how to refute them. We soon found one probable basis on page 296 of the first book: a Babkin claim that I told him I worked for the CIA.

The books consisted of the collected protocols compiled by the FSB, mostly from interviews with me, but also, I now saw, from interviews with Kiely, Babkin, and others at Bauman and at Region; with contractor Bolshov and his son-in-law; and with a dozen others who had participated in meetings with me. I was intrigued to learn that the "evidence" did not include interviews with many of my Russian contacts whom the interrogators had earlier said they'd debriefed. Some volumes contained documents, others did not, so I had to flip back and forth to read the supposed evidence against me. Also included were my medical evaluations by the prison doctors, Andrey Andrusenko's credentials and certification as my lawyer, and similar claptrap in quantities that overwhelmed the few documents of real importance. Two crucial contracts were nowhere to be found in the string-bound pages: the agreement between Russian Technologies, Region, and myself to pursue commercialization of marine technologies and Bauman's certification of the Babkin reports as containing no state secrets.

I was permitted to look at only one or two books at a time, in concert with one of the five different interpreters brought in for this purpose. One of the new interpreters actually laughed at the absurdity of the "evidence." In another room, Andrusenko from time to time examined and took notes on one or two volumes that I was not then using. We were not permitted to go over the books together, although occasionally, when we met, we did review notes. It was as though we were separately preparing to defend the same case.

In the thousands of pages, there was plenty of material that disproved the FSB's charges. For instance, on page 61 of Book One was a protocol in which Babkin admitted using material from his dissertation in compiling the papers for ARL — a dissertation that had been stamped "secret" — but asserted that the extracts that he used did not contain any state secrets. In Book Two, Myandin, who had worked with Babkin, asserted in three separate interviews spread over three months that the materials on the HRG were not secret, and no secrets had been passed to me; Myandin also gave the interrogators chapter and verse on preparations by himself and others on the Bauman/Region team, made prior to meetings with us and designed to make sure that they and we adhered to the proper security guidelines. Book Eight contained the texts of articles, such as one by a principal designer of the Shkval, in the March 1996 issue of *Military Parade,* touting the sale of Shkvals to other countries and concluding, "Russia is ready to offer to carry out mutual research and design work on separate disciplines and creation of various versions of high-speed underwater rockets."* There was also the reprint of a

*Genrikh Uvarov, "Underwater Rockets: Myth or Reality?" *Military Parade,* March–April 1996.

Region brochure advertising the Shkval and a copy of a 1996 letter from the Committee on State Military Technology Policy giving Region permission to work with "foreign orders" on the technological bases of the Shkval.

Most of the bad news was in Kiely's protocol. It had him saying that he and ARL did work for the Army, the Navy, Lockheed-Martin, and other defense contractors and that Pope probably had been under orders to find and steal "closed military activities." Kiely's materials also contained some weird stuff, in reports that had nothing to do with me but which certainly appeared suspicious, such as a project summary that included a "confusion scenario" in which either Viagra was to be put in the drinking water or a news report was promulgated to that effect in order to wreak havoc with the enemy. From Kiely's protocol, I learned for the first time that very late in the evening on April 3, he had signed a "nondisclosure agreement" with the FSB. What was Kiely not to disclose? Did the FSB really think Kiely would honor this document? Another volume had the notice that on August 7 — just days before I was given the books to read — all pending charges against Kiely had been dropped.

Despite the almost-welcome work of going over the books, I spent a lot of time in my cell staving off boredom, inactivity, and paranoia. At times being cooped up with three, four, or five men was difficult, because by this time I had made a conscious decision not to learn more of the Russian language; I reasoned that to study it so I could better converse with my cell mates would mean accepting the idea that I would remain in a Russian jail for a long time, which I refused to do. But we cell mates did communicate on various levels.

During this late summer period, the most vivid personalities in the cell were Sergey and Viktor. Sergey was a huge man, six feet five, 300 to 350 pounds, who was being detained on suspicion of tax fraud. He spoke English well, and whenever I would appear on television, say, during a court appearance, and at many other times when the TV news mentioned something about the United States, he would stand to attention and sing "God Bless America" in English at the top of his lungs. Sergey would continue bawling though the guards would bang on the cell door and command him to quiet down. He also upset the guards' routine: when they pointed at you through the privacy partition in the cell door, you were supposed to recite your name, last name first; when they pointed to Sergey, just to annoy them he would reply, "Putin, Vladimir Vladimirovich."

Viktor was a nervous, imperious chain-smoker who paced incessantly, drank buckets of coffee, played a lot of backgammon, spent hours on end outside the cell, and had considerable influence with the prison administration, which led me and other cell mates to conclude that he was an agent of the FSB. Whenever Viktor became annoyed with a cell mate, he'd announce that he would have that man removed from the cell, and, sure enough, within a day or two that removal would take place. Viktor was probably responsible for having Sergey removed from our cell after only a short residence. Viktor did not like to watch me on TV and would change the channel during the news broadcasts; I'd change it back, arguing that the television set was mine. Usually he'd give in on that point.

Although we all had occasional disagreements, and my cell mates did regularly steal my medications — primarily to see if they could be used to get high — kindness, courtesy, and humor were practiced every day among the detainees in cell 73–74. To my cell mates, the most interesting delicacy that I received was Fig Newtons; they had never tasted anything like these cookies and tried to steal them when I was out of the cell. As a result I was forced to eat the remaining Fig Newtons more quickly than originally planned.

I learned from the guys how to prepare salads in the Russian manner, by dicing vegetables into very small pieces, adding salt and mayonnaise, and infusing virtually every concoction with minced garlic.*

*Recipe for "Salad Sergey": half-quart grated carrots, 4 ounces grated Parmesan cheese, mayonnaise to taste, and fresh-killed Lefortovo mosquito.

That summer, we participated in occasional periods of silliness; I was surprised and charmed, one day, to realize that several cell mates were acting out routines made famous by the Three Stooges, and I joined the fun. A hand held perpendicular to the face and between the eyes became our salute. We called the cell "The Cave" and gave each other nicknames keyed to the animals that lived in such an environment. Viktor was the Danger Fox, Sergey the Danger Bear, and I was the Danger Spider — the moniker given to me, in part, because in Russian the name Pope and the word for spider sound somewhat the same. To be called a spider in Russia verged on a term of endearment; spiders are thought of as successfully crafty and as good luck totems, not as pests to be killed.

The nickname sat comfortably, because after more than four months in a Russian jail I had become fairly sly — not an unreasonable response to living within a web of deceit. I had learned how to make requests of the prison administration, how to have the guards shuttle materials back and forth from my cell to the lockers, how to share small luxuries with my cell mates, and also how to hide things that I preferred not to have them pilfer. One of those was a small Bible, two by four inches, that my Dad and Mom had sent over; my parents had presented it to me in 1960, but I had left it at home at some point during the intervening years. I was happy to have it in Lefortovo and kept it in my pocket, partly for inspiration, partly to prevent cell mates from pinching it.

Our jailers allowed us to have only two photographs apiece in the cell. Several men had extensive families whose multiple photos they wanted to display; two photos per man was a needless cruelty. A prohibition against feeding birds outside the window was another. When one of us was about to be taken out of the cell for an interrogation session, he would chew and swallow a clove of raw garlic so his bad breath would annoy the interrogator. For the few cell mates who were obviously FSB plants, voicing disgust at the FSB meant having to do very credible acting jobs.

Most of the cell mates were sympathetic and believed in my innocence, but a few decided that I must be a spy. For the least intelligent of the latter bunch, I put on a little show, based on the FSB not having removed my ring. When the cell window was open, I held a metallic spoon to the ring and pointed it in the direction of the sun, as though the jury-rigged device was a sender and receiver of radio signals. The next day, the guards executed a full search of my body and belongings, but found nothing incriminating.

In the United States, my family, friends, and supporters were bringing pressure on the handful of people in the upper levels of the government who could effect my release. The very realistic fear that I might die in prison succeeded in changing the attitude at the State Department from lackadaisical to animated, if not yet to zealous. Had a second American citizen died in Russian captivity, public opinion in the United States might have forced some high-level resignations at State. Strobe Talbot, Ambassador Pickering, and Secretary Albright, separately and in concert, began to bring up the subject of Ed Pope repeatedly in their many interchanges with Russian officials. Publicly and privately, they assured the Russians that I had not been working for the U.S. government — something that State had not said explicitly prior to August. A State Department spokesman said in a briefing to reporters, "We have made clear [to the Russians] that they bear the responsibility for protecting the welfare of American citizens detained in Russia."

This was all very welcome, but my family and I could only wonder why the same tone had not been adopted a hundred days earlier, because in the interim the facts of my case had not changed. One difficulty with State's new approach was its narrow focus, taking the form of requests to have American doctors evaluate my health. Not only did this not address my innocence, but the Russians read this as criticism of their own medical system and resisted efforts to have me independently examined by, for instance, an embassy doctor — an American accredited diplomat who was to have visited me along with Brad and June Kunsman, another embassy official who shortly replaced Brad as my regular visitor.

The State Department action that seemed to carry more weight with the Russians, in mid-August, was a threat to issue a "travel advisory" for those going to Russia for business. But this, too, did not result in my release. And by the third week in August, when Congressman Peterson had enough cosponsors to assure the passage of his "Sense of the House" bill — thanks to my supporters from all over the country, who importuned their representatives on my behalf — he was advised by the House leaders, Pavel Astakhov, and people at State to postpone a vote. The thinking was that if a request for a release on bail or for medical purposes arrived in the Russian court simultaneously with the passage of a congressional resolution demanding that I be returned home and threatening consequences if I was not, the Russians would be in no mood to grant the relief sought in court.

On August 24, I was handcuffed and taken in a special bus with its windows blacked over to another location in Moscow for a hearing of what was termed the General Prosecutor's Preliminary Trial. As I was being shuttled to the courtroom, I was startled by a clutch of reporters and cameras and a shouted question — in English — from a man I later learned was Dave Montgomery of Knight Ridder. He asked how I was doing. "Okay," I told him, but also said that I could not comment on the charges against me.*

*Montgomery graciously e-mailed Keith McClellan with the news of my sighting, as he knew my family and friends would like to have a firsthand report. He said that I looked "tolerably okay," but that the disappointment at not being released was visible on my face. Montgomery continued to pass information in a timely manner throughout my ordeal, his phone calls to Cheri on various events frequently beating those of the State Department to her on the same matters.

It was the first time I had seen a Russian courtroom, and during the proceedings I was held in a large cage inside the room. During a brief hearing, Pavel Astakhov argued that I should be released on bail and sent to a hospital for the purposes of medical treatment prior to my scheduled trial. This would have been a perfect opportunity for the Russians to relinquish custody of me — as some people had told me they wished to do — but that was not to be. After a brief deliberation, the judge denied the bail request.

However, he did grant me permission to call my dying father in Oregon. Pavel considered this a minor victory, and told the press outside that he feared the phone call would remain just a promise.

I suspect that the entire hearing had been requested only to provide the news media with a good "photo op," and Pavel with a way to use the media coverage to warn the FSB that this case would not be decided wholly behind closed doors or adjudicated only in a court of law. The court of public opinion would also be its venue, and Pavel, a well-known star performer, would use the public stage to the advantage of his client and himself. In his press conference, for instance — which I watched that evening on television — he made the point that the seized documents, which the FSB had claimed were classified, were really unclassified, and that it was the FSB, rather than his client, who had first revealed their details to the press, thus breaking their own rules on discussing the case in public.

Twelve letters from home were delivered to me that afternoon upon my return from the hearing. I devoured them. One was a card that had been sent by Cheri in June. It featured an angel, and she wanted me to keep it by my heart as a reminder of her; I tried to do so by secreting a portion of the card in a book. When the time came for me to give back the mail, I turned in the dozen letters. The guard walked off with them, and five minutes later came back to search the cell for the missing part of the card. I feigned innocence, but they found it. "Oh, I forgot. I used it as a bookmark," I claimed. They seized it and told me to write an apology to the warden for having kept it, or else. I wrote one that any native English-speaker would have recognized as so overblown that it had to be a parody.

The report of my court appearance and the denial of my release request hit the U.S. media on August 25. That day, as well, my former boss, retired Rear Admiral Tom Brooks, was informed by the *Washington Post* that his letter to the editor about me would appear on the 27th. When I eventually got a look at it, I wondered why it had been so tame, and Brooks told me how it came to be printed in that form. Around the first of August, he had written the piece and sent it to the *Post*. At that time, its most incisive paragraphs read:

> One would assume that the U.S. government has been sparing no effort to get this American businessman released. . . . But little appears to have been done. It would seem that the administration is more concerned with "not offending the Russians" than it is with the life of a U.S. citizen and former naval officer who served this country well for over 25 years.
>
> It is time for the Congress and public to speak out and insist that the Clinton administration, so apparently seized with human rights in some of the most remote areas of the globe, remember its primary responsibility to protect the rights of its own citizens traveling abroad. It is time for Congress to take a no-nonsense stand with regard to the Russians: treat American citizens properly or forget any aid or assistance from the United States.

He heard nothing from the newspaper for several weeks; then, the newspaper's staff made changes that would considerably soften the letter's anger at the apparent inaction from the Clinton White House and State Department. In the past, Brooks had written letters to the editor and op-ed pieces for several newspapers, including the *Post,* but had never before been subjected to editing for content, only for style. Someone — most likely in the White House or at the State Department — appeared to have convinced the *Post* that in regard to the Edmond Pope case, the paper should not print the sentiments of a former Director of Naval Intelligence without censoring them. Brooks did not like the changes and beat some back, but felt at last that he had no choice but to accept most of them or else nothing would be printed. By the time his letter was published, it had been shorn of direct criticism of the administration. There was, for instance, no sentence about an administration "more concerned with 'not offending the Russians.'" Instead, it read:

> Mr. Pope's wife has met with President Clinton. Members of Congress from his home state of Pennsylvania have gathered more than 140 sponsors for a House resolution to put pressure on the Russians to release Mr. Pope, whose health is reported to be failing badly. Mr. Clinton twice has spoken with Russian President Vladimir Putin [news story, Aug 17], but Mr. Pope still languishes in prison without adequate medical care.
>
> Congress and the public should insist that the Clinton Administration remember its responsibility to protect the rights of its citizens traveling abroad. It is time for Congress to take a no-nonsense stand with regard to the Russians: Treat American citizens properly or forget any aid or assistance from the United States.

The *Post*'s introduction to the letter also did not mention the quite relevant fact that Brooks was a former Director of Naval Intelligence, which would have given greater credence to his words.*

My lawyers told me that Cheri would be visiting me again on August 29, and this time would be accompanied to Russia by Congressman Peterson as well as by his press aide, Jen Bennett.

*Brooks had not mentioned that fact either in his letter, but if the *Post* was checking the letter with government authorities, it could easily have learned about and put in this important reference.

The circumstances of this visit differed so markedly from those of Cheri's earlier trip that it was as though someone had flipped the embassy light switch from Off to On, from Arm's-Length to Helpful. Jen Bennett had already scheduled interviews for Cheri and Peterson with CNN, *Good Morning America, The O'Reilly Factor* on Fox, and the ABC newsmagazine *20/20*. With such coverage imminent, the State Department could no longer be cavalier in its treatment of Cheri or her congressional escort. When the Peterson party arrived at their hotel, they found lists from June Kunsman of phone numbers and contacts to call in case of difficulties. On the earlier visit, Brad Johnson had refused to provide Cheri and Jen Bennett with his direct line. The embassy staff set up meetings requested by Peterson with members of the Duma and with the American business community in Moscow. Peterson understood that the new, gracious treatment by the embassy merited a quid pro quo, that he and Cheri not openly criticize State for past inaction. Peterson's refusal to do so had caused him to part company with Congressman Weldon, a perennial critic of the State Department who had frequently used the inaction of State in regard to my case as an opportunity to take a shot at the Clinton administration; Peterson, though as consistently opposed to Clinton's policies as Weldon, reasoned that skewering State or Clinton would not serve his main goal, to get me out of Russia.

Back in the U.S., my advisers were urging their friends and colleagues to watch these American television broadcasts from Moscow and to respond by sending messages to elected representatives asking them to sign on to Peterson's bill, and to demand that the State Department issue the travel advisory.

For Cheri, the television and radio appearances were awful but necessary. "She is a very private person," Jen Bennett later recalled, "she prefers to be behind the scenes, and hated doing the interviews, even though she did them really well. This is not a woman who wanted her fifteen minutes of fame; all she wanted was to get her husband back, and then slip out of public view."

The toll that the past five months had taken on Cheri was apparent as soon as I saw her in the Italianate visitors' lounge of the prison, and the sight hurt me. As for Cheri's impressions of me, she would shortly tell reporters that I was ravaged, had lost even more weight, seemed frail, had alarming-looking lesions on my face and neck, was shaky and weak, and complained of near-continual pain at the base of my skull. Our hour alone turned into two, but the time still flew by. I had brought into the visiting room a page of notes, most of them about aspects of my defense and my health problems; we went over them. She spoke of having given to the prison authorities for me the first two volumes of J. R. R. Tolkien's trilogy, *Lord of the Rings;* we had read the trilogy thirty years ago, when first married; now, she said, I was to read the first two volumes, and we pledged that when I got out, we'd read the third one together. So involved were we with conveying pent-up feelings without dissolving into tears that we both forgot that I was supposed to pass her a hair from my head, so it could be brought back to the U.S. and analyzed for poisons and to provide information on my medical condition.

After the visit, Cheri went outside to face the cameras. Responding to a CNN question, she said, "It has been a very difficult day for me. Having to leave him was extremely difficult. I have to . . . call his mother and father and our sons and tell them that Ed is not doing very well, and when he does come home it's going to take a very long time to get back the son and husband and father that left our house in March." The day became more difficult and hectic for her, although very productive, with a half-hour live television interview with Bill O'Reilly, during which the Fox network put on the screen the phone and other contact numbers for the White House and the State Department. Other prominent television programs also broadcast interviews with Cheri and Peterson. Switchboards at the White House and the State Department soon lit up and stayed lit for days with calls demanding more government action toward my release. Cheri and Peterson also appeared on the most popular radio call-in show in Russia, *Echo Moskvy,* where Clinton had previously been a guest. Questions from the in-studio interviewers started out alleging that I had been involved with sensitive technologies, but soon became more sympathetic, as did questions and comments from the ordinary people who called in.

A day after Cheri's visit to the prison, the State Department in a regular briefing let it be known that they were about to issue the long-contemplated travel advisory. In reaction to this harder line, the FSB released to the Russian press, for the first time, word that Dan Kiely had been detained in connection with the Pope case. A radio report said that the FSB had seized documents that proved both of our espionage, but (as the *Moscow Times* put it), had then "released [Kiely] due to his senior age and willingness to testify."

If in early April I had been set free and Dan Kiely held, then upon returning home I would have immediately started a ruckus and would not have piped down until my colleague was released. But when Kiely returned alone he did not do that, probably on instructions from ARL. The laboratory deputy director, Liszka, had told Cheri very firmly, "We don't know what Ed was doing over there in Russia," and Penn State had put out a statement that the university did not feel that my arrest was an issue for the university to become involved with. I have since learned that employees of ARL were also instructed not to have any contact with members of my family, or to assist them in any way. Today, I am convinced that if at that critical juncture in early April, Kiely, Liszka, Hettche, ARL, and Penn State had all come out in vociferous defense of me — as any friend or neighbor would have done — American public opinion could have been marshaled behind them, helping the White House put pressure on the Russians to withdraw their "espionage" case, and I would have been returned home within a very short period of time.

Instead, I was left to rot in prison while Penn State and ARL tried to distance themselves from me. They gave various excuses for not assisting me, such as that Keith had asked them to refrain from public comment. Not so, Keith says. Hettche even refused to respond to legitimate journalistic inquiries about my employment at ARL, or what I had done with and for ARL, such as my part in siting the Ukrainian electron beam program at Penn State, or the Penn State seminars with Russians that I'd organized — facts that could have helped effect my release. According to Brad Mooney, who had known Hettche and Kiely for years, when he met privately with them, it appeared to Brad that Hettche refused to look him in the eye while discussing what had happened to me.

The intransigence of ARL was tested on September 1 when the FSB publicly revealed that Kiely had been detained and had given what the FSB considered sworn testimony. According to the FSB, Pope and Kiely were a team: Pope had arranged for the secret purchase of data and Kiely, the associate director of ARL, had assessed the relevance and value of the information. Therefore, the FSB contended, "All the talk from the Americans about how [Pope] acted alone, and was not linked to the gathering of intelligence, is total nonsense. They [the Americans] should question their professor Kiely, who was sent back expressly for such activities."

The American weekly newspaper *Chronicle of Higher Education* picked up the Kiely Moscow story and requested a comment from Penn State ARL. It came from the vice-president of research at Penn State, who said, "All of Dr. Kiely's visits, and Dr. Kiely's collaborative research facilitated by Mr. Pope, were approved by high-ranking Russian officials. Charges suggesting that Dr. Kiely was involved in intelligence are completely without merit."*

And what about charges involving me? Penn State and ARL continued to be silent on this issue.

My family and friends were enormously upset at this latest ARL sin of omission. Keith charged into the fray, reminding the ARL people how their silence had damaged me, and suggesting precisely what positive assistance they could have offered had they felt like

*Bryon MacWilliams, "Russian Authorities Say Penn State Researcher Participated in Spy Operations," *Chronicle of Higher Education*, September 5, 2000.

it. Later in the month, as Keith and Peterson's office were preparing documents for Pavel Astakhov, Keith asked ARL for clarification regarding my contracts, and provided Hettche and ARL with another opportunity to proclaim my innocence. Their response, which included the idea that ARL had been in contact with the Russians only through me — that is, not directly — further distanced them from my activities and made no reference to my innocence.

Both Cheri and I were members of the Penn State community. Ray Hettche had personally recruited me to work at ARL, and Kiely's work with the Russians, which had begun before I joined ARL, had been facilitated by the connections I had made for him in Russia. Why would Hettche and Penn State choose to disavow me? One possible answer came to me while in Lefortovo: upon learning that the charges against me were based on documents carried by Kiely, it occurred to me that Kiely and other employees of Penn State and ARL might have been instructed to remain silent because the university, and certainly ARL, might have feared they had some liability for negligence in regard to the hurts I suffered, since ARL had permitted its employee, Kiely, to take those documents with him after I had expressly warned Hettche's deputy that Dan should not do so. Interestingly, when Brad Mooney later asked Kiely about the contents of the March 17 phone call, in a closed-door session Kiely managed not to mention to Brad that I had told him to stay home if he was worried about his safety, or that I had instructed him to leave sensitive papers behind.

In early September I fell ill once again. I feared it was the cancer recurring; I had been told in the U.S. that my immune system was weak and easily compromised, and these current illnesses seemed proof of that. They also made me afraid that my severe mental depression would again engulf me. But when I asked the prison authorities for simple remedies — a laxative, a sleeping pill, mild pain-relievers — I got the runaround. After a week of this, I wrote a strong note to Shelkov, which ended with the words "I consider your actions that have put me in the present situation as a direct and serious provocation." My cell mates told me that to accuse a Russian of provocation was a slap in the face; in this instance, the irritant produced results. Shelkov angrily told me that my health was not in his purview, and that henceforth I should address the prison administration, not the interrogation team. But almost instantly, and surely because Shelkov yelled at them, the administrators directed their clinic to pay slightly more attention to my complaints. They would still not deliver the bulk of the medications that my U.S. doctors had sent, which had been passed to the prison authorities by the embassy. They did bring in several "specialists" to examine me: an oncologist who seemed incompetent and also claimed that she was currently treating several cases of my very rare cancer, which could not have been true since there were very few cases in the world; a dermatologist who belittled the growths on my neck, face, and nose; and an ophthalmologist to treat my headaches, on the presumption that they were caused by changes in eyesight. The prison authorities went to great lengths to prevent me from being taken to a hospital or an independent clinic for evaluation and treatment. My cell mates and I agreed that all the specialists I had seen were, at best, proctologists-in-training.

On September 8, news broadcasts reported another spy scandal: a Russian naval attaché had been detained in Tokyo and then deported for buying from a Japanese naval officer classified documents concerning the propulsion system of a new torpedo. The Japanese officer had confessed. Thinking in crafty spider terms, and looking for positive omens, I wondered if my former colleagues in the Navy had helped set up that arrest to highlight that Russia was accusing me, a former naval attaché, of stealing what Russia's own naval attaché had just tried to steal from Japan. There are all sorts of ways to send signals.

Later that evening — it was actually reported the following day in Russia — President Putin appeared for ten minutes on *Larry King*

Live on CNN. Russian newspapers called King the toughest interviewer Putin had ever faced. Actually, King was more polite than searching; when he asked Putin what had happened to the *Kursk*, Putin coyly answered, "It sank," and King did not follow up with another question about the disaster. However, and to my joy, King did ask Putin very directly about my case. To a question about the matter proceeding to trial, Putin responded, "In our country, like elsewhere, the legal process should be finalized and then, depending on the situation and certainly in the spirit of good relationships between our two countries, we'll see what we can do." That certainly sounded hopeful for me, since it strongly implied he might release me after the inevitable conviction. "Even if a court confirms that Mr. Pope has caused some substantial harm by his activities, I don't really think that intelligence can be that harmful," he said, and to another question about my health, he responded, "If it comes to the situation when it's up to me to make the decision, then naturally this will be taken into consideration." That, too, was a strong hint that I might be released once the trial had concluded.

When I saw the clips from the interview I reveled in hope, but my optimism had to be tempered, because I am a believer in Yogi Berra's maxim "It ain't over till it's over." A deal might well be in the works, but at least one such deal had already gone sour; a thousand things could go wrong before the moment for my release arrived, and any one of them could change the circumstances, or Putin's mind, or his ability to overrule a guilty verdict.

Andrey and I were rushing to complete our notes on the evidence; the books were to be taken from us shortly and given to the judges to study for a month prior to the trial. This seemed absurd to me, because a copy machine could have easily been used to provide all sides in the case with copies of the relevant documents. But the lack of copy machines, a feature of Soviet life that had been used to cut off the spread of anti-Soviet information and literature, still controlled the pace of justice in Russia.

On September 14, June Kunsman from the American embassy visited me, bearing the unnerving news that Cheri was talking about moving to Moscow so that she could come and see me every two weeks; I asked June to plead with Cheri not to do that.

On September 19, I was taken to the clinic and blood samples were drawn; there had been little sleep the night before, so when I returned to my cell I lay down for a while, only to be interrupted by the guards summoning me for a second journey to a court. This time, I was taken in the bread truck, handcuffed inside a small cage. The destination was the Moscow City Court.

"Mister Pope, is that you?" asked a voice in Russian-accented English coming from a nearby cage in the truck. Startled, for a moment I made no answer, so he asked again and this time I responded with a wary "Yes." The speaker identified himself as Valentin Moiseyev, and wondered if I had heard of his case. When I professed ignorance, he told me about it. Moiseyev had served in the Ministry of Foreign Affairs for many years, in various posts, including a stint in North Korea and one in South Korea. He had also visited the United States. In 1998 he had been arrested, tried, and convicted of selling Russian state secrets to the South Koreans between 1992 and 1998. The FSB said that a South Korean diplomat had been caught with sensitive papers given him by Moiseyev; Valentin maintained that the sensitive document was the text of a public lecture he had given. The conviction was overturned by a higher court on the grounds that the lower court had been too vague about precisely what secrets had been purloined. A new trial had been ordered, and he had been kept in Lefortovo while awaiting its start in September. As his wife had told the press, "It's clear that the verdict has been programmed and this trial will be another farce." Putin had even said publicly that Moiseyev's guilt had been proved beyond a doubt, well before the new trial had begun.

Later that day, I was able to see Moiseyev in person, though we could not shake hands because of our handcuffs. I spoke to him several more times during the daily trips back and forth to the Moscow

City Court, where his trial was already in progress. He was a gentle, erudite man, increasingly ravaged by his years in detention; we were almost the same age but he appeared a dozen years older.

My court hearing was again perfunctory. The Moscow City Court denied Pavel's plea to have me moved from prison to a hospital, pending trial, on the grounds that the offense with which I was charged was too serious to remove me from Lefortovo. This time, however, reporters got close enough to ask questions. I was able to assert that I was not receiving proper medical care, and that "the nature of my cancer means that the longer I am here [in prison], the greater the danger of a problem."

The process of reviewing the evidence books came to a halt. Everything else stopped, too. The clinic aides who had recently been looking in on me three times a day did not show up; my medications were held back; and when June arrived for a regular consular visit, she was turned away. I did not see my lawyers for three weeks; they also had been turned away on various pretexts. Mail forwarded for me was held by the warden's office and not delivered; mail that I tried to send was returned to me with a note that the postage was not sufficient; supplies such as envelopes, which I had ordered and paid for, and which I had been told had arrived, were likewise not delivered. My written requests for obtaining reading books and other materials from storage went unheeded. It was clear to me that I was being put on ice, held in isolation until the trial could begin.

After a month of waiting for the Russians to make good on the judge's order to allow me to phone my dying father, on September 25 I pressed the prison authorities to facilitate the telephone call. The warden's office at first said they didn't know how to do it; then they said I'd have to pay for it, which I instantly agreed to do; then the request sat overnight, and the next morning I was informed that it had been summarily denied. The episode was a grim reminder that in Russia the FSB/KGB is more powerful than a court order.

* * *

Testifying before the International Relations Committee of the House of Representatives at the end of September, Secretary Albright characterized the Russian handling of my case as "outrageous," the strongest publicly expressed sentiment yet by an American official. Aware of Peterson's resolution, which was before the committee, Albright cautioned the representatives not to make a "foe" out of Russia even if they were angry over the treatment of Ed Pope. On October 5, the IR Committee unanimously approved the bill, which recommended that the president tie future Russian aid to my release; the full House would take it up within days.

My first news of the House committee's action came from a new source — the pages of the *Moscow Times*, an English-language daily, which, after several months of waiting, was after October 1 being delivered to me by mail five days a week. It was a blessing to be able to read pertinent and objectively reported information on a daily basis. The next day, the news was even better: the State Department finally issued the official travel warning:

> In Russia certain activities that would be normal business activities in the United States and other countries are still either illegal or are considered suspect. Americans should be particularly aware of potential risks involved in any commercial activities with the Russian industrial-military complex, including research institutes, design bureaus and production facilities. Any misunderstanding or dispute in such transactions can attract the involvement of the security services and lead to investigation or prosecution for espionage.

These were very strong words coming from the State Department. The Russian Ministry of Foreign Affairs and the Duma reacted angrily, contending that the warning was an attempt to exert pressure on the Russian judicial system in my case. The MFA put out a statement saying that "according to documents which appeared in court, Pope was involved in the gathering of secret information. . . . It is scarcely likely that this could be described as 'usual' business activity."

Any pleasure I derived from the MFA's whines disappeared on October 6, when I was taken to a reading cell and shown my formal indictment. It was now six months since I had been arrested, and this was the first time that my lawyers or I had actually seen the indictment. In less than two weeks, the trial was scheduled to begin.

The first stunning fact was that the indictment contained no charges having to do with the bulk of the work I had been doing in Russia: there was nothing about Pustovoit and his acousto-optical tunable filter, which I had been publicly accused of stealing, nothing about my various endeavors with the Krylov Institute, nothing stemming from my trips to Novosibirsk or Kaluga — no charge, that is to say, based on *any* of the papers in the "A" stack seized in my room and which had ostensibly provided the grounds for my arrest.* The indictment was twenty-two pages of balderdash based almost solely on the five Babkin reports that Kiely had brought, spiced with misreadings of my background in Naval Intelligence, and topped by my supposed links to the "military-industrial complex," as evidenced by the few business cards from representatives of Lockheed-Martin and several other companies that I had in my possession.

Laboriously, I penned "corrections" to the indictment, running to forty-nine sections totaling dozens of pages. Since the indictment itself became part of the court record, and I was never given a copy to keep — because it was ostensibly a classified document — it can only be partially reconstructed through my notes of corrections. In section #16, a typical one, I took exception to a paragraph charging that I "persuaded [Babkin] to reveal classified information . . . and . . .

*The reason the FSB dropped charges referring to those endeavors, I believe, was that interrogation of the Russians involved had revealed that many of those lateral, non-Bauman enterprises also concerned my acquisition of work on drag-reduction and other technologies related to supercavitation propulsion, a fact that would have buttressed my assertion about the innocuousness of the Babkin-Bauman work.

to collect espionage information." My correction said, in part, "If Babkin ever revealed any classified information, it would have been done in contradiction to our intent." I began to work up these "clarifications" and "corrections" into an opening statement for the trial — the supposed start date, the 18th, looming ever closer.

On October 9, the NTV news program *Criminal Journal* broadcast an incredibly relevant story. The manager of an aluminum plant in an outlying region had been brought into Lefortovo for questioning about several murders in the Moscow area that he was accused of arranging. But investigative journalists for a magazine associated with NTV had now discovered that the whole affair was a frame-up by the FSB. None of the bodies of the supposed victims could be found, and there were many other problems with the investigation, all traceable to the mendacity of the investigating agency; it appeared that the FSB's objective had been to get back at the aluminum plant manager, probably for refusing to cut them (or their Mafia partners) in on some deal.

The next day, October 10, in the culmination of months-long efforts by Congressman Peterson, my family, and supporters, the multisponsored HR 364, retitled HR 404, was passed by the full House of Representatives in a voice vote. This nonbinding resolution called for my immediate release on humanitarian grounds, and urged that until I was released, the U.S. ought to provide no funds for the Russian Federation to join the World Trade Organization, and that "the President should use the voice and vote of the United States to oppose further loans to the Government of the Russian Federation by any international funding institution of which the United States is a member." The White House promptly expressed its pleasure at the passage of this resolution, saying it strengthened the hand of the State Department and administration officials who were now trying to effect my release. Russian reaction was predictable. The Speaker of the Duma said that the resolution amounted to "crude interference in our criminal procedural legislation." A liberal who was deputy head of the Duma's defense committee called the resolution "a great stupidity" and opined that Clinton would probably ignore it.

I was very grateful for the outpouring of sentiment and the marshaling of force that HR 404 represented; it was greatly encouraging for me to know that the entire U.S. House of Representatives had taken a firm stand on my behalf. However, as I prepared to go on trial for espionage in Moscow on October 18, 2000, I wrote in my notebook, "After all these six months I still find it very difficult to accept that I am in the middle, actually the subject, of such a crazy incident; however, as each day goes by, it becomes more evident that my captors could have told me in April what day the trial would start, and possibly to the day how long it will last, and what the verdict/outcome will be, as if we didn't know that already."

CHAPTER 9

No Themis in This Court

On Wednesday, October 18, the long-awaited trial began.

In a routine that would be followed for the duration of the proceedings, in the morning after breakfast at Lefortovo I was put in the bread truck, handcuffed inside a cage, and taken on a ten-minute ride to the court building. Since the truck had no windows in the rear section, I could not see the exterior of the building or be certain of its location. The truck pulled into a small inner courtyard, and I was taken out and escorted down a corridor with its own stairwell, a back one, used in order to deny the waiting news media and other onlookers any glimpse of my arrival at court. I was walked into the basement, then up two flights of stairs to the second floor; from there I was brought into courtroom 227 before anyone else was present, and locked inside the defendant's cage.

Room 227 of the Moscow City Court was a setting devoid of majesty. A rectangle perhaps thirty feet wide by sixty feet long, its main unusual feature was the defendant's cage, located on one side of the room and measuring about ten feet by twenty feet, with bars from floor to ceiling. Inside the cage were benches that could have held two dozen people squashed together; in all my life, I had never felt lonelier. Having such a contraption in a courtroom perfectly mirrored a system that considered the defendant guilty until proven innocent. During the trial, when I was speaking or being addressed, I was required to stand in the cage, and the frequency with which I was addressed made it necessary for me to stand for hours at a time — an indignity that was also a form of torture.

Next to the cage were my two attorneys' desks, and slightly forward of them, and to the other side of the rectangle, was the prosecutor's desk. Three guards were present at all times. A translator — my old "friend" Alyosha, from the interrogations — was stationed near my cage, closer to me than my attorneys were, and he roamed up and back to confer with the judges and prosecutor. The three judges in their black robes sat on a raised platform at one end of the rectangle; they usually entered the courtroom from their chambers behind the bench. Directly in front of them sat a court reporter with a computer keyboard and other materials. Room 227 was a minimalist setting, deliberately drab, with the same rotting wooden parquet floors that were standard in all Soviet-era buildings.

In the opening week of the trial, I remained handcuffed until I had been brought into the defendant's cage, but as the trial wore on, the guards removed the handcuffs as soon as I got out of the transportation van and permitted me to walk unfettered into and out of the courtroom. Occasionally, cameras would catch sight of me down a long corridor on the second floor, when the door to the courtroom was opened to admit my attorneys or other functionaries of the court. I wore my sports coat, tie, and business shirt for these court appearances, and in some photos that I saw of myself in that attire, I appeared a lot healthier than I felt.

At the back of room 227, opposite the judges' dais, were rows of seats that in other trials might have held family members and observers; during my "closed" trial they were kept empty. The ostensible reason for this being a closed trial was that the material under discussion contained state secrets, but actually the Russians just did not want any media or outside observers (such as officials from the American consulate) having firsthand knowledge with which to criticize the proceedings. Emblematic of the secrecy surrounding the trial, the news media were never able to report the chief judge's name properly; they mistakenly kept referring to her as Nina Barkina, because they had no access to court documents correctly identifying her as Nina Barkova. They weren't the only ones in the dark. Her name had been kept from me and my lawyers until the first morning of the trial; Pavel, who had wide experience in that court jurisdiction, did not know of her even by reputation. She was in her late forties or early fifties, a formidable person, of small and athletic build, with iron-gray, close-cropped hair and somewhat masculine facial features. For a while, we were uncertain if she was male or female, as her voice was low and harsh. She called everyone "comrade," and was to be addressed as "comrade." The other two judges were an older man and an older woman; they had perfected the trick of sleeping with their eyes open, and during the trial seldom said a word or displayed interest in the proceedings.

Americans are so used to a justice system founded on the bedrock notion that an accused person is innocent until proven guilty that we are almost unable to comprehend all the concomitants of a judicial system that only pays lip service to that notion. For example, we take for granted that separation is to be maintained between the operations of the prosecutors and those of the judges; but in Moscow City Court, the three-judge panel and the prosecution acted as a tag-team, with the judges asking questions of the witnesses that the prosecutors might well have asked. Another feature of an American court is frequent conferences between the defendant and his lawyers; my lawyers and I were not allowed to consult on many matters. For instance, they were not permitted to advise me whether or not to answer a specific question; nor could they raise objections to the testimony of witnesses, to the questions asked of witnesses by the prosecutors or the judges, or to unfounded allegations made against me by the prosecutors. At any time the prosecutors or judges could ask me, the defendant, a question, and I would have to answer it; moreover, such a question would not be followed by one from my own attorneys, the response to which could serve to put my earlier answer into context. The judge's and prosecutor's constant querying of me, and their continual attempts to turn innocuous matters into serious allegations, eventually drove me to cynicism. When they asked baldly if CERF and TechSource were among the American companies who took orders directly from the CIA, I responded, "Of course. As you know, *all* American companies take orders from the CIA." After that retort failed to get a laugh, I had to explain, "That's a joke. As you know, I'm no James Bond." The joke and the reference irritated Comrade Nina.

Prosecutor Oleg Plotnikov was a tall, slim, and stern senior officer from the prosecutor general's office; he appeared to be doing his job, though with relish for drawing nefarious inferences from every piece of evidence. He also had a predilection for picking fights with Pavel, whom he hated, calling him "Mr. TV" or "Mr. Hollywood," and railing at him for representing Gusinsky. The two of them would even argue over how much money was involved in Gusinsky's rapacity. But my case was hardly in Plotnikov's hands to prosecute. Comrade Barkova routinely asked most of the obvious questions of the witnesses before he got in his licks, and did so in a manner so prosecutorial that he hardly had anything left to ask.

I was continually amazed at the lengths to which the judge and prosecutor would go to prevent any hint of fairness creeping into the proceedings, as though just a smidgen of truth would cause the downfall of the edifice of lies they were so laboriously erecting. The tenor of the court was made blazingly clear to me on the first day by the judge's handling of four key motions made by my attorneys.

Motion one was to get rid of the FSB translator, Alyosha, and replace him with a translator independent of the FSB and of the prosecution. Pavel, who spoke English reasonably well, would often have to correct Alyosha's translations of the proceedings into English for me, or correct Alyosha's translations into Russian of my English answers to questions. As Pavel later explained to reporters, "A person who is subordinated to, or, because of his work, dependent on the leaders of the investigation cannot but be at least indirectly interested in the success of his department." Alyosha would hang around me, overhear my conversations in English with Pavel and Andrey, and then go and report them to the judges and prosecutor. He would frequently convey notes from outside the courtroom to the prosecutor, and in other ways act as though he was an official of the FSB who outranked the state prosecutor. Later on during the trial, when the men of the "expert commission" came to testify, we could see Alyosha working hard to keep their testimony in line with what the FSB wanted them to say, and watching them closely so that he could report to his superiors if the experts did not do as they were told. Once in a while I was able to use Alyosha's tattling for our benefit, feeding him misleading information that I knew he'd convey to the prosecutor and judge.

Our second motion: if the translator could not be replaced, we wanted to have a video and/or audio record of the proceedings made, so that we might later ascertain whether the translation had been accurate or biased.

The third motion dealt with perhaps the most prejudicial matter in the case, the absence of Anatoly Babkin from the prosecution's list of witnesses. I believed that were Babkin to be questioned on the stand, he would exonerate me. But the prosecutor and Comrade Nina were determined to prevent that from happening. They wanted to have introduced only a statement that Babkin had given, on paper and on videotape, back in April, when the old professor had first been detained. The prosecution contended that Babkin was now too ill to testify in person. His case had been severed from mine, and he would be charged and tried at some later time, the prosecution said. Pavel argued that to deny us the right to question Babkin in person was prejudicial.

The fourth motion dealt with the absence from the evidence books of the permission letter from Bauman allowing us to commission and obtain the Babkin papers.

On Friday, October 20, the second day of the trial, Comrade Barkova denied all four motions. The impact on my defense was devastating: there would be no impartial translator; the proceedings would not be recorded by audio or video; Bauman would not be subpoenaed to provide a copy of its letter of permission to explore the HRG technology; and Anatoly Babkin, the key witness, would not be called in person to testify.

In the United States, each of these matters, by itself, would have been enough to overturn a court's verdict on appeal. The Russian judges' rulings on these four motions made it obvious that not only would there be no attempt at fairness in this trial, but that even the semblance of fairness would vanish. By the trial's end, six weeks hence, Judge Barkova would have denied more than two hundred defense motions, many of them similarly regarding key documents or testimony that would have exonerated me in any Western court proceeding. One, for instance, asked the court to give to the defense a copy of the apparent governmental basis for charging me with espionage, which had been elucidated in two presidential decrees and a government resolution. Barkova ruled that these three documents were secret, so we could not see them or have them introduced into the court proceedings.

From the *Moscow Times* of October 25, 2000, I learned more about secrets in Russia. According to staff writer Anna Badkhen, thirty-six different government ministries were permitted to compile their own lists of secret information, even though some of the lists had been found to contradict extant laws and to violate provisions of the Russian constitution. It was also against the law to base espionage charges on secret documents; an edict to that effect had been used in 1999 to acquit Alexander Nikitin, the navy captain accused of espionage for assisting environmentalists. But Nikitin's case had been in St. Petersburg, and mine was Moscow, so perhaps the law did not apply countrywide.*

*Anna Badkhen, "Defining Spying Is Murky Business," *Moscow Times,* October 25, 2000.

* * *

In an American court, opening statements are the province of lawyers; in this Russian court, it seemed expected of me to conduct part of my own defense, and so, along with Pavel, Andrey, and the prosecutor, I also made an opening statement. I welcomed the opportunity, because it seemed the only way to get on record my objections and corrections to the protocols — an opportunity that had been promised to me many times, but which had never been given concrete form. Using the indictment as a guide, I tried to answer its many allegations, starting with the Russian assertion that I had been working for the CIA, which I flatly denied, and proceeding on to the multiple misstatements about what I had been doing in Russia for the past nine years. In the statement, I stressed the step-by-step nature of our approach to Region, Bauman, Russian Technologies, Rosvoorouzhenie, and Professor Babkin, and the careful ways we had evolved to have Babkin's reports reviewed by security committees before they were sent to us. "None of the reports provide detail which would be sufficient to build any hardware or to manufacture the fuel [for the Shkval]," I said. I also took a shot at the head of the committee of experts that the FSB had appointed to review the Babkin reports and had declared they were classified, telling the court, in regard to Logvinovich, the head of that body, that "his past activities provide an anomaly at best and perhaps a very clear case of conflict of interest; because of his past role, his committee findings are very suspect."

That was too much for Comrade Judge; she waved her arms in the air and declared she had heard enough and ordered me to stop speaking. But I got in one final statement: in conclusion, I stressed that we had come to such entities as Region, Russian Technologies, and Bauman by invitation, had proceeded in a "slow and deliberate manner to ensure that our Russian partners would have time to adequately complete security requirements," but in return for these efforts to follow the rules, "we have been cheated, lied to, tricked, and persecuted." That last line earned me the first of Judge Nina's evil-eye stares.

Later that day, when Pavel left the courtroom to talk to reporters about my opening statement, he characteristically added a flourish of his own and attributed it to me. He said that I had asserted that in this particular trial, Themis, the goddess of justice, was absent from the courtroom. Later, I learned that in Greek mythology, Themis is the daughter of Uranus and Gaea; she is the only immortal (aside from Zeus) who did not fight for one side or the other during the Trojan War, an action that led to her being considered impartial, and to her becoming the goddess of justice. In his classical allusion, Pavel hit the nail on the head: the contrast between impartial Themis and the completely biased Comrade Judge Nina was complete.

Not surprisingly, Comrade Nina came to hate Pavel, and for good reason. Pavel's conversations with the waiting news media, each day after the trial session, let the cat out of the bag in terms of what was happening inside her courtroom. His comments were also vastly important to the conduct of my case and to the shift in public opinion that took place while the trial held the headlines for six weeks. Since no one else from the courtroom was speaking to the media — rules forbade doing so during a "closed" trial — Pavel was able to control what the outside world learned of the trial. By showing up the heavy-handedness and ineptitude of the FSB, the prosecutor, and the judge, Pavel countered Putin's desire to conduct a show trial that would reflect well on the FSB and the courts.

At times Pavel would exaggerate things for our benefit, both his and mine. For instance, he told the media that if our motions for an independent translator and for recording devices were turned down, his client had decided that he would remain silent for the duration of the trial. I had made no such vow, although I had considered the idea. Since the beginning of the affair I had always answered questions, based on my belief that an innocent man must not be evasive or choose silence; having good and legitimate explanations for all of my actions, I saw no need to clam up. But during the first portion of the trial it served Pavel's purposes — and, likely, mine — for him to threaten the court with my refusal to testify; when I read in the *Moscow Times* that I would not testify, I chuckled, but did not chide Pavel for misrepresentation. Later on in the trial, when I did answer questions in the court, Pavel reported to the press that I had relented in order to hasten the trial along; that scored a few more points with the public on my behalf.

Pavel also claimed to have other weapons to bring to bear on my case. His wife, Svetlana, had been a friend of Putin's wife since childhood, and so Pavel had what was in effect a back channel to the president of the Russian Federation. Whether he actually used it on my behalf, I did not know. He claimed to send messages to Putin in other ways; for instance, on one of the rare occasions when Putin was at a location remote from Moscow and taking questions from the press, Pavel said he arranged for local reporters to ask Putin about me; the importance of that moment, according to Pavel, was to demonstrate to Putin that he would not be able to avoid scrutiny about, and responsibility for, my case.

Based on my limited knowledge of the American criminal justice system, I had expected the prosecution here to present its case, with time for cross-examination of prosecution witnesses by the defense, and then for the defense to present its side, with its own witnesses, and also with opportunity for cross-examination of our witnesses by the prosecution. I was entirely wrong. After opening statements, the trial consisted of the judge and the prosecution going through the evidence books from beginning to end, and calling witnesses to testify along the lines of what they had already been quoted as saying in those books. The defense might cross-examine these witnesses, but would not have the opportunity to present its own case until everything in the evidence books had been hashed over. Most of the trial was therefore taken up by either the judge reading the books aloud, or by the judge or prosecutor questioning me about the protocols. Rather than taking the stand to endure a concentrated but limited period of questioning, I was in effect on the stand all the time, except for short periods when someone else was testifying, and even during a witness's testimony I would be asked questions by the tag-team of prosecutor and judge.

Occasionally, when Pavel and I wanted to keep our tactics secret, or just to annoy the prosecution, we spoke in Swedish. This would produce consternation among the FSB translator, the prosecutor, and the judges, as though a completely private conversation between the defendant and his lawyer was something outrageous or illegal.

In the middle of the second week of the trial I caught a glimpse of Cheri in the hallway outside the courtroom, framed by a forest of cameras. I had been told that she was coming, so I was not stunned. The Russians would not allow her into the courtroom. That day she stood outside the court as the trial proceeded and refused to heed the guards' requests to sit down; so long as I had to stand in the cage, she told them, she would stand in the corridor.

They would also only allow her to visit me for an hour back at Lefortovo, after which she would not be permitted to visit me for another entire month. During this meeting I was full of aches and pains, and because of the trial and its unfairness more nervous than I had ever been in my life. The time with her was balm for my soul. Her most important message was this: "Some friends," she said, "want to know what you would like to eat for Thanksgiving dinner. And these same friends have asked me to tell you that you are going to have to buy me a very expensive Christmas present."

This message excited and confused me. Did it mean I would be home for Thanksgiving, or that the embassy would bring me Thanksgiving dinner in prison? Did it mean that I would definitely be out of Russia in time for Christmas? Was the message based on a deal having been made, or only on hope? Cheri also told me, as I had already heard from June, that she wanted to come to Moscow to live, an idea that undercut the promise of my being released before

Christmas. I told her to wait on that decision until the conclusion of the trial.

She also conveyed distress over Pavel's grandstanding. She believed he was acting more in his interests than in mine, and that he relished her presence in Moscow because it increased the number of cameras and reporters before whom he could perform. She thought he was slowing the trial down to have more opportunities to make the nightly news. On that score, at least, Cheri was wrong, because it was Judge Nina who controlled the tempo of the trial.

Cheri's prison visit was brief. Later, I learned that Cheri had been so upset at my mental condition that she could not bring herself to stay for the entire hour. "My big strong husband has been reduced to a nervous wreck," she told Peterson; "they've broken him. It was all he could do during that hour to keep from crying."

Fortunately for me, John Peterson was running for reelection unopposed, and he seized the opportunity to devote his time and energy to my cause. By then, Peterson and his office were receiving daily reports on the progress of the trial from Pavel, bulletins every other day from June Kunsman at the embassy, and frequent updates from Tom Pickering and other high State Department officials. His intervention activities were even waking the dead — at Penn State. Prior to Peterson's journeying to Moscow in October, the university's president, Graham Spanier, sent the congressman a large batch of papers regarding my ARL employment, ARL contracts with Bauman, details of the arrangements with Babkin, and even a university statement regarding my innocence. "This would have made all the difference in April," Peterson told Spanier. "Then, this information might have freed Ed; but now that the trial has begun it would only distract from the main issue, the need to bring Ed home before he dies in prison." Accordingly, Peterson left the materials at home.

The televised appearances of Cheri, Peterson, and Pavel in Moscow greatly aided my case. Cheri's bald statement that her husband was in bad health "and will die if he is not released" led the news on every channel. When Peterson was refused entry to the prison during Cheri's visit to me there, he used the time and occasion to speak at length with the assembled press. They were impressed that an American congressman would put himself out for an American citizen; no member of the Duma, the Russian press told Peterson, would have done that for a Russian constituent. Reinforcing the concern of the Congress for this case, Peterson was able to emphasize that on the day of the trial's opening, October 18, the U.S. Senate had unanimously passed the companion legislation to HR 404, putting the Senate, too, on record as also demanding that I be released and threatening to cut off American assistance to the Russian Federation if I was not.

It was Peterson's office that had coordinated Cheri's visit, and that, too, served multiple purposes. Cheri's warmth, her concern for me, and the passion with which she imbued her roles as wife and mother touched many people in Russia. Old women in head scarves would come up to her on the street and express their admiration and sympathy. A Russian journalist apologized to her "for what this country has put you through." What Cheri, Peterson, and Pavel told the press about the trial, the charges, and my mistreatment was soon echoed by the increasing number of comments from the State Department and the White House. We received word that on the 20th Clinton had telephoned Putin to complain that he had not yet done enough to effect my release; according to the White House, President Putin responded that he could do nothing now because the case was in the hands of the court.

Tangible evidence of the changed attitude about me in the Russian press could be seen and heard in the words they used about me. Until the onset of the trial, Russian press reports had called me "the American spy," but two weeks into the trial, after finally receiving massive doses of information about my view of the arrest and trial, even such state-dominated enterprises as RIA-Novosti began to refer to me as "an American businessman accused of spying." Polls taken in Russia showed that fewer than half the respondents now believed that I was guilty of espionage.

Unbeknownst to me, another thing John Peterson did while in Moscow on this trip was get tougher with the State Department. The congressman told Ambassador James Collins that I might die if things dragged on too long, and demanded that Collins immediately obtain assurances from the Putin government that Ed Pope would be released at the end of the trial. John's insistence served to spur the cautious ambassador to action. Within the day, word came back to Peterson that Collins had relayed the congressman's message to Washington, that President Clinton had taken this new message directly to Putin by telephone, and that Clinton had been personally assured by Putin that at the end of the trial, the Russian president would "use his presidential powers" to have me released. John Peterson recalls being only marginally relieved by these assurances, for there were still many things that could go wrong before Putin's latest promise was redeemed.

When a witness would take the stand, Comrade Barkova would question him first; when she was through, Plotnikov would take over. Then it would be Pavel's turn, and after him Andrey, and then I would have the opportunity to ask questions of the witness. Routinely, many of Pavel's and Andrey's questions, and an even higher fraction of mine, would be disallowed. "It is not authorized," Comrade Nina would say of a question, or "I will not allow that." Of course these were the questions whose answers would obviously be favorable to our side. Barkova's rulings on their admissibility seemed entirely arbitrary; moreover, she never provided reasons for disallowing a particular question. Pavel would argue with her on these rulings, but after a few interchanges she would tell him to sit down and shut up. Nor was there any appeal to a higher court in regard to particular questions or on such matters as what witnesses to call or documents to subpoena; the only thing that could be appealed, in the Russian judicial system, was the final verdict.

The first substantive witness at the trial testified on the day Cheri was in the corridor. Arsenty Myandin, a propulsion engineer, worked for Region and was one of the lead designers of the Shkval. As Myandin took the stand, I was quite apprehensive, because although my minimal dealings with him had all been aboveboard, I did not know if he or any of the other Russian witnesses might have been compromised by the FSB and forced to testify. Myandin's testimony, as elicited by judge, prosecutor, my attorneys, and me, showed how much bias the FSB had put into the protocols derived from its interviews. His testimony on the stand differed so dramatically from the protocol, it was revelatory. He insisted that the technology that I had sought was not secret; that I had obtained the correct permissions to allow it to be transferred; and that he had actually written some of the Babkin papers — this was news to me, as I had thought Babkin had written the reports himself — based on lectures that Babkin had been delivering openly for fifteen years. Myandin said that no one would be able to duplicate the Shkval with the information in the five Babkin reports, and also corroborated my assertion that these reports had all been subjected to review and approval by a Bauman University security committee before they had been sent to the U.S.

Myandin was an entirely credible witness, and very valuable in terms of the information that he gave to the court: that Shkvals had been sold on the open market to other countries, that the HRG technology already had many civilian uses, that I had always stressed to himself and Babkin and Bolshov that we were not to cross over the line into discussing aspects of various technologies that might remain secret. He also said that the research he and Babkin had done for the reports was based entirely on open sources.

"*Very Positive!!!*" I noted exultantly in my notebook.

In a Western court, Myandin's testimony would have produced a Perry Mason moment, in which a witness's testimony so undermines the prosecution's case that it collapses. But not in Moscow City

Court, where Judge Barkova simply dismissed Myandin after he was finished testifying, and went on with the case as though nothing of importance had taken place on the witness stand.

The next witness was Gennadi Pavlikhin, vice rector of Bauman MSTU. His appearance before the tribunal was noted in the press, and Pavlikhin in court made a formal protest: he had not wanted his name known to the general public. Pavlikhin's testimony echoed Myandin's, explaining for example, that I had been the one who had insisted that reports from Bauman or Babkin must not contain any classified material.

Similar gladdening testimony then came from other witnesses who had had dealings with me, such as Tatyana Danilenko, Alex Maliarovsky, and Yevgeny Shakhidzhanov, the director of Region. I noted that in Maliarovsky's testimony he had talked about "work with L-M [Lockheed-Martin] thinking this to be positive. I now know the goons will view this as negative." In fact, I had done no work with or for Lockheed-Martin, although I'd hoped to involve them later on with the HRG project, so I asked my lawyers to introduce a letter from that company which expressed their very preliminary and non-binding "interest" in the technology, should it become available. Judge Barkova refused to allow the letter into the court record. Why? Because, she said, it was in English.

In contrast to most other witnesses, Shakhidzhanov came out and faced the cameras after testifying. He told the press, "I described my relations with Mr. Pope, who was interested in high-speed underwater movement. Our institution [Region] did not hand over any classified information to Pope, but this sphere has classified elements." Interfax, a Russian news agency, reported in connection with the Region director's testimony that the Shkval torpedo had been completely declassified, but cited prosecutor Plotnikov as saying that while technical data about the Shkval had been advertised in a military magazine, the technology for producing the torpedo and its fuel were still secret. Of course that was not an accurate representation of what Shakhidzhanov had said; in my notes was a direct quote from his testimony as translated by Alyosha: "Even if you buy fuel you cannot understand how to produce."

Two of Pavel's most frequently repeated requests, denied time after time, were for the live testimony of Anatoly Babkin and the obtaining from Bauman University of its committee's "security review" documents. Pavel had actually seen these "security review" documents — a friend inside the FSB had shown them to him, he told me — but the FSB had refused to enter them into evidence. Withholding such evidence in a Western judicial system is not only malfeasance that could bring about an immediate mistrial and/or dismissal of charges against the defendant, but is itself a crime for which errant prosecutors have from time to time been charged. In the Russian system, withholding exculpatory evidence seems to be standard prosecutorial practice. In this instance, however, it appeared so blatantly unfair — a condition brought about by the publicity attendant on the case — that after Pavel's fourth or fifth request on the subjects, Comrade Barkova suddenly decided to have the documents brought in, and later to call Babkin to testify.

The Bauman security review documents arrived, were introduced as evidence, Pavel and Andrey made motions for dismissal of the case based on these documents, those motions were denied, and the trial continued on as before. The judge still refused to call Babkin to the stand.

On October 31, a Tuesday, during the morning session of the trial I had a lot of pain in my back, neck, and elsewhere, and nearly passed out in the cage. As was usual, I was taken back to Lefortovo for lunch, and there suffered excruciating pain, so debilitating that I refused to go back for the afternoon session and insisted on seeing the doctor. This time, my illness alarmed the medical personnel at the prison enough so that they ordered me to have at least two days' rest before continuing on with the trial. Their diagnosis was acute rheumatic inflammation of the back and hip joints. I thought it might be far worse than that, possibly a recurrence of the bone-based cancer that had effectively ended my naval career.

In court, Pavel Astakhov was so vehement about my illness that Judge Barkova suspended the proceedings for two days; outside the courtroom, Pavel briefed reporters on the unusual development, taking the opportunity to tell the public that Cheri had several times hand-carried to Russia the appropriate medications for my conditions, but that the prison authorities had refused to give them to me because those particular medicines were not registered in the Russian Federation. The news of my acute illness reached Washington. I was later told that because of my deepening illness President Clinton again called President Putin, and Secretary Albright called Foreign Minister Ivanov, to ask them to do something. Prosecutor Plotnikov commented to reporters, "The mere fact that the U.S. Congress, Secretary of State Albright, and Clinton himself pay so much attention to Pope makes one doubt he is just a humble owner of a small company employing three or four people. . . . I think he is an operating intelligence officer and to get him off the hook direct pressure is being applied on the court."

The immediate result of the direct pressure was that for the first time in six months of being detained by the FSB, I was taken to a real hospital, inside the Kremlin, for an MRI and for a consultation with a neurologist. The Kremlin hospital visit was a relief to me. I was treated professionally and courteously at a place obviously reserved for VIPs. Because I was being brought to such a place that day, the Lefortovo guards — dressed for the occasion in civilian clothes — treated me in a more friendly manner. While I waited for the results of the tests, a hospital official showed me an article from the *Kommersant*'s reportage on my case, which quoted Shakhidzhanov saying that I had not obtained secret information from his outfit. I autographed the article and added my thanks for the care and hospitality extended to me. He took the paper, beamed, and put it in a drawer in his desk. The MRI results were ready, and the prison doctor discussed them with the Kremlin doctors. When I asked about the results,

Alyosha told me that everything appeared normal "for my age": there was no evidence of cancer but there were signs of osteoporosis. But since they would not show me the films, I could not entirely trust this diagnosis. Osteoporosis was an entirely new thing with me, likely the consequence of having been imprisoned for six months, with terrible nutrition, minimal exercise, and maximal stress and tension on my body; furthermore, osteoporosis is a bone disease that could trigger my hemangiopericytoma. At the close of my hospital visit, the chief administrator, who spoke some English, walked me back toward the prison transportation van. In a quiet moment, he gave me a double-handed grasp and had tears in his eyes as he assured me that this ordeal would end and that I would soon be going home. I wanted very much to believe him.

Perhaps he knew more than I gave him credit for. Unknown to me at the time, on November 2, my mother in Oregon received a telephone call from the Russian ambassador to the United States in response to a letter she had sent to Putin; the ambassador said he was calling on behalf of Putin and conveyed the following message: there was nothing that Putin could do while the trial was being conducted, but that after it ended, he would find a way to send me home. (Cheri had sent three letters to Putin, but never received a call from the ambassador.)

In Lefortovo, of course, I knew nothing of such assurances, and if I had, I would no longer have put much credence in them. Then, while I was "resting" for two days, Judge Barkova decided for reasons then unknown to me to suspend the trial "indefinitely."

Over the weekend, the reason for her suspension of the trial became apparent: there was a huge revelation in the making, one that could throw the trial into chaos.

It began with a television program. NTV ran an audiotape that it characterized as having been received "under rather complicated circumstances." The tape featured two men speaking with Anatoly Babkin and his wife, Galina Babkina, in the Babkin apartment. According to later transcripts released by NTV, the two men, who identified themselves only as Oleg and Andrey, say they have come to visit them "unofficially . . . because some of the nuances do not satisfy us." The nuances were in the confession signed by Babkin, which he had also read for a video camera. The visitors instructed the couple that when Babkin took the stand during the Pope trial, "Anatoly Ivanovich must say only what he said during the investigation. Anatoly Ivanovich, if you suddenly say something other than you have, you will get Article 275. . . . Remember, the most important thing: you must say only what you said in the investigation."

Article 275 is the Russian statute related to treason. The rest of the conversation with Oleg and Andrey made their offer very clear: if Babkin repeated at the trial what he had confessed to during the initial interrogation that began in April, the charges against him would be changed to those under Article 283, which dealt with disclosure of state secrets, a lesser charge that carried a lesser penalty, and which would be eligible for an amnesty.

In a separate, on-camera appearance, Babkina repeated the thrust of the surreptitiously taped conversation in her living room, and asserted that her husband had been forced to sign a confession in April, fearing that if he did not, his heart problems would go untreated and he would die in Lefortovo. They held him anyway, from April to August, and only then, after signing and recording a second confession, had he been permitted to go home.

On November 9, a letter from Babkin arrived in court. It read, in part:

> I recant my testimony concerning Edmond Pope, which I gave under pressure, having a grave heart condition preceding a heart attack. I signed the protocol of my interrogation without even reading it. In the same condition and also under pressure, I read my testimony before the video camera. In fact, I have never met Edmond Pope face to face [alone] and have not given

> him any sensitive information. Edmond Pope himself has never asked me to give any kind of sensitive data. Herewith I am asking (the court) to consider my testimony concerning Edmond Pope to be not corresponding to reality.

The judge tried hard not to deal with this stunning development — the government's major witness recanting his earlier testimony. Pavel first tried to deliver the letter himself, but she would not allow it. Then, it came in properly, as a "protocol" sent directly to the court by Babkin and accompanied by other documents such as a doctor's letter attesting to what Babkin's condition had been at the time of his arrest and detention. Barkova put off a decision, but she knew that the press had received a copy of Babkin's recantation letter, so she could not avoid the matter entirely.

By this date, the presidential election in the United States had been held, but the winner was in doubt. I had hoped that as soon as the American election was decided, Putin would release me as a gesture of goodwill to the newly elected president, whoever that might be. When the winner was not officially declared and the post-election contest between Vice President Gore and Governor George W. Bush commenced, my heart sank, because I now had to assume that no relief would come for me until the outcome of that battle was determined. But then came word from Judge Barkova that Anatoly Babkin was to be called as a live witness. Freedom of the press had forced her to do something she would not otherwise have done. In order to appear as not giving in to the defense, Barkova simultaneously ruled that the members of the state "experts" commission that had judged Babkin's papers as containing classified material ought to sit in and listen to his testimony, before they gave theirs.

We all expected a moment of high drama as the seventy-year-old professor shuffled in, looking very frail but dignified. I recalled that on the day we had been arrested, I had been on the verge of telling Anatoly Babkin that we were going to cut back on any further arrangements with him. Now, here he was, a man who had had the courage to stand up to the FSB and the entire Russian judicial system on my behalf.

Babkin took the stand and began to say in answer to Barkova's questions everything I had longed to hear: that since he spoke only Russian and I spoke only English, we had never conversed out of the hearing of a translator; that his earlier testimony to the FSB had been false; that he particularly regretted signing a document that said he knew that I was an agent; that he had never given me secret material, that his reports had been cleared by the Bauman censors before being sent to the U.S., and that those reports were based on open sources. He had even brought a stack of the open-source textbooks with him to enter into the court record; of course, Judge Barkova would not permit that. I was surprised to learn from Babkin's testimony that he had based his earliest work on HRG on a technology licensed from the AeroJet General Corporation of the United States in 1962. If true, this meant that I was on trial in Russia for stealing a technology on behalf of the U.S. that had originally been developed in the U.S.

As Babkin testified, he also kept clutching his heart, stopping the proceedings so he could take pills for his condition, and in other ways performing as though he were a very bad actor imitating a sick old man. Several times he gave the same answers to diametrically opposed questions; he contradicted himself right and left; he tripped over his words; he appeared dim-witted; he forgot certain matters, some for his own convenience, others out of befuddlement. He waltzed into a very recognizable prosecutorial trap when he was pushed to the point of admitting that he said that in our discussions he had never used the name "Shkval," an assertion that was easily refuted by other evidence. He was also seemingly confused about his own five technical reports. "Comrade Babkin, did you write this report?" "No, Myandin wrote it all." "Well, did you obtain this photograph?" "Yes, that is from an article I wrote in 1964, which I copied for — well, maybe I did write some of the report, after all."

The prosecutor began to laugh; the experts laughed; Judge Barkova laughed; even Pavel and Andrey had to chortle, and so did I, at the ludicrousness of the situation. But since I knew that Babkin was basically telling the truth in his testimony, and that truth was all I could count on to exonerate me, the laughter was full of pain. This day had perhaps been my best chance at an acquittal. Now, I felt, there was nothing that could stop them from convicting me.

CHAPTER 10

Breaking the *Parasha*

Babkin's testimony and its rejection by the court pushed me into a severe depression. It was made worse by the testimony given and the exhibits shown in the next few days of the trial.

The first blow was a surprise witness. Normally, when my lawyers would introduce a motion to have someone testify, it was refused; once in a while such a motion was granted, but we'd have to wait a week or ten days for the witness to show up in court. But when the prosecution unexpectedly asked permission from the court to call Oleg N. Kosichkin — a person unknown to me or to my lawyers — the request was casually granted, and a moment later Kosichkin waltzed in to testify.

He was billed as the head of the Bauman MSTU technical review commission and was a rebuttal witness to the testimony of Pavlikhin, who had stated quite forcefully that the review commission had vetted all the Babkin reports before they were sent out of the country. I had never seen Kosichkin's name on a document or a list, but he testified that he was the leader of an eight-person security-review commission, and that each article or report was usually reviewed by only two or three of the members. He revealed that there was no record-keeping or tracking system; they did not even keep copies of their approval letters. He couldn't say what had happened to most of the Babkin papers, except for one, the first, because he had participated in that review and had vetted it positively as containing no classified material. The implication was that the other four had not been vetted.

Then it was the turn of the expert commission to be witnesses. In the first court session, Georgy Logvinovich and Yuri Fadeyev (general designer for Region) answered softball questions from Judge Nina and Comrade Plotnikov and reiterated the commission's findings that the Babkin papers had contained classified material about the Shkval. This finding was at the heart of the government's case, so we needed to attack it.

It was my habit, when I knew who might be testifying in the next court session, to go over the notes I had taken while reviewing the protocols, and to use these as a basis for drafting questions. By this time I had half of a six-man cell to myself, while my two roommates occupied the other half, which permitted me more room and privacy to go over my notes without intruding on anyone else's space. Since Comrade Barkova wanted all questions for the expert commission submitted in advance and in writing, I prepared in my cell more than two dozen for Logvinovich and Fadeyev. Reading the English translation of the commission's report had made it obvious to me that the report had been written by the FSB, not by scientific experts. In many parts and phrases, the wording of the report was identical to that in the summary of the indictment, and the vocabulary used was not that of scientists but of Soviet-era hacks. Accordingly, the questions I wanted to pose were designed to demonstrate to the court that I knew, and that the court should acknowledge, that the FSB had drafted the report. Pavel and Andrey also prepared dozens more questions. All our questions were submitted to the judges, and after a two-hour deliberation Comrade Nina refused to allow twenty-seven of my twenty-eight and permitted only five or six composed by my lawyers.

These queries were passed to the commission members, who were then given time to write out the answers in a venue away from the courtroom. We could guess who had actually written their responses. Still, the Logvinovich and Fadeyev answers to Pavel's questions revealed that the three absent members of the commission might not even have seen the Babkin reports before rendering their judgments. In answer to a question by Andrey — echoing the opinion of Pavlikhin, who had vetted the Babkin reports for Bauman, that the experts on the commission were not qualified to make a decision regarding the classification of the *propulsion* system of the Shkval — Logvinovich had to admit that his main field was the body of the torpedo, and that none of the experts on the panel had any expertise in security matters.*

Eventually, two of the three missing commission members showed up in the courtroom. One of them was a Navy captain and engineer. They answered a few pro forma questions from the prosecutors; after the session, Fadeyev stepped outside the courtroom and declared to newsmen that the commission had determined that the data in the Babkin reports "has never been declassified and we are not going to declassify it."

Another pillar of the prosecution's case was the surreptitiously recorded videotape of a meeting in my room at the Sayani on March 18 between myself, Babkin, Bolshov, and Titov. Excerpts from this tape had been shown on television in response to NTV's airing of the Babkina audiotape — and, in excerpted form, the videotape did appear somewhat incriminating. The excerpts could be construed as showing us conspiring not to tell other people what we were doing, and to work out ingenious ways in which I would give Babkin money that he was not to share with others. Here are two excerpts:

TITOV [in Russian, translating a previous line from Pope]: We need to agree and be absolutely clear who can talk with whom, and who can't.

*During a recess, Logvinovich and I chatted; I apologized for having accused him of once visiting the United States, when he hadn't done so, and he accepted that apology and went on to tell me that he had done a great deal of work outside of Russia, in Italy and France; he was about to expand further when Alyosha yanked him away and told him to shut his mouth.

POPE [in English]: I don't want him to get in trouble —
TITOV [in Russian]: Ed doesn't want you to have problems.*
POPE [in English]: — and I don't want to be in trouble, either.

In another section of the videotape that had been featured on television, while Babkin was out of the hotel room for a moment, I spoke about him to Bolshov.

POPE [in English]: What about this problem with the payment? I've had this problem with Babkin before: he conveniently forgets.
TITOV [in English, replying after consultation with Bolshov]: I remember, but I wasn't involved at that time. Do you remember how he lost a thousand [dollars]?
POPE [in English]: Yes.

Having seen these excerpts played over and over on television, I dreaded what the FSB would do with them in court. But there was a mild surprise, because instead of just the excerpts, the whole tape was played. And in its entirety, the tape made the contexts of the excerpted portions much clearer — and more favorable to me.

To begin the session in my hotel room, I reviewed the long sequence of events that had led us from the Ministry of Defense to Region, to Bauman, and thence to Babkin. Babkin then talked about the letter of authority that had enabled his work but said he didn't want to tell Shakhidzhanov, the director of Region, what he was currently doing. I raised this as a question, "Should we tell him . . . ?" and Babkin answered that just now we should not, and also that he, Babkin, should not discuss his work with Russian Technologies. Given this context, the first excerpted portion, which was pulled from my end-of-the-meeting summary of the discussion, could be

*According to a native-Russian speaker in the United States, who later listened to this conversation as reproduced in a Russian "documentary," Titov's formulation in Russian of my words to Babkin was not as clear nor as direct as my statement in English.

seen in a new light: I had been making the point that we in the room had agreed with Babkin's insistence that no one other than the contracting parties should be told about the work he was *currently* doing — current work that did not form part of my indictment.

The second excerpt concerned a side payment of one thousand dollars that I had made to Babkin, and his request — seen and heard loud and clear on the videotape played in the courtroom — for more money for his projects. After hearing his demands on that day, I told Babkin he must make requests for payment and submit all bills to Bolshov, because I had forwarded to Bolshov's company, for disbursement, all the money that was to be paid to contractors. Babkin protested and tried to make it seem as though he was still owed the thousand dollars that we both knew I'd already given him.

A third excerpt played on television concerned the type of fuel used by the hydroreactive generator of the Shkval. The four of us in the hotel room had had some difficulties in discussing this, because we understood that part of the information about the fuels was classified. The videotape had Titov translating into English a Babkin comment that "on Shkval, we had a lot of trouble with this aspect." Again, the full-length context was edifying. On the whole videotape, the court could see and hear Babkin bring up the idea of my buying some interesting fire-fighting equipment that used the same HRG and fuel as the Shkval; the fire-fighting equipment was not classified at all, and so we would be able to acquire the fuels information in a roundabout way. Hearing this from Titov, in translation from Babkin, I responded in my summary very firmly, "There will be no discussion or money paid for a contract having to do with the fire-fighting system."

The trial had gone on for about a month, and by this point there was, finally, a lot of communication taking place among my attorneys, the American embassy, Peterson's office, other parts of the State Department, and my family. The team on my side was firing on all cylinders.

Ambassador Pickering had called my mother and Cheri, and from wherever he was on the globe he would daily telephone Peterson with an update on my case. These calls were tremendously comforting to all the recipients, evidence as they were of concern for me at very high levels of the American government.

Imprisoned in Lefortovo, I knew very little about these communications, and consequently — and fortunately — remained unaware of a long-distance battle being waged between Cheri and Pavel. Repeatedly, Astakhov had sent word through Peterson's office, and through June Kunsman at the embassy, requesting that Peterson and Cheri come over to assist him at the trial. Since there was nothing that either of them was permitted to do inside the court, their presence in Moscow would have meant only more rounds of television and radio appearances. These would undoubtedly have put some additional pressure on the FSB, the court, and Putin — but likely not enough to move the Russians forward at a faster pace. Cheri remained unconvinced that her presence in Moscow just then would do me any good and refused to fly over again unless assured that the trial was about to end and that I was on the verge of being released. Pavel could not give her such an assurance in November, but wanted her there anyway. She demurred, telling John Peterson that when the trial showed signs of wrapping up, she would go to Russia, but until then, she did not want to suffer the indignities that the Moscow authorities would heap on her

Had I known of Cheri's stance at the time, I would have agreed with it, though I missed her enormously. During this mid-trial period I found it increasingly difficult, as I told my diary, to "keep my chin up." Weekends were the most challenging for me, because even though I tried to exercise every day, and to distract myself with chores such as laundry and reviewing notes on the protocols, I found myself staring at Cheri's picture. I wrote her a note, though there was no hope of delivering it: "I miss you so much it hurts. Several times I have broken down and gone into tears. I never realized just how much I really do love you; it gets so bad at times, it physically hurts."

In court, day after day, Andrey, Pavel, or I would introduce motions, virtually all of which would be refused by Judge Nina, often with a smirk on her face as she looked directly at me. One of our most important motions was a demand that Bolshov be called as a witness; he had been in the room during the hidden-recorder videotaped session and could testify about it and about Babkin's participation; also, he was my partner in various enterprises and would be able to back up many of my contentions. He would also have been able to testify that in our conversations at the Sayani on April 3, before Babkin had arrived, I had made a decision to move the entire HRG demonstration-model project from Babkin, Bauman, and Region to Russian Technologies because I was dissatisfied with the current work being done by Babkin and Region. The FSB did not want Bolshov on the stand, and so Judge Nina would not permit him to testify. We submitted dozens of other similarly relevant motions seeking the testimony of witnesses who would corroborate my contentions.

Looking back now on the denial of these motions, I have to wonder: What were the FSB, the prosecutor's office, and Judge Nina afraid of? If the trial's verdict was fixed in advance, and the process was just for show, why not let in anything the defense wanted? My best guess is that to do so would have influenced the view of the case in the court of public opinion, for there was no holding back Pavel when a camera or a microphone was pointed in his direction, and if he had a scrap of evidence to go on — say, the testimony of a Bolshov — he would play it to the hilt and with it dominate that evening's televised news and the headlines in the next morning's newspapers.

After one too many such exhibitions of Pavel's ability to influence the media, and after Plotnikov had made repeated threats to Pavel Astakhov along the lines of we-know-what-you're-doing-and-you-won't-get-away-with-it, one morning Plotnikov arrived in court with a sheaf of what he identified as transcripts of Pavel's conversations on the case, drawn from a tap on Pavel's phones. There could not have been much of substance from the taps, since Pavel had assumed his phones were compromised anyway; once, to avoid being overheard, Pavel took John Peterson into the middle of a traffic island in Moscow to convey information about his strategy in the trial. In a Western court, to have admitted tapping the phone of the opposing lawyer would of course have provoked a mistrial, but not here; Judge Barkova didn't bat an eye at the flaunting of the fruits of the tap in her courtroom.

In mid-November, Grigory Pasko published a book about his experiences of recently being on trial for treason. Pasko was a former Russian naval officer who had been charged with that crime for giving to Japanese journalists evidence that the Russian fleet was dumping toxic wastes into the seas near Japan. In 1999, after an enormous international outcry, especially among environmentalists, Pasko was acquitted of treason by a military court but convicted of the lesser charge of abuse of his office. He appealed the conviction, and in the interim before the higher court ruled he wrote his exposé of the trial, of the Russian crackdown on environmentalists, and of the failures of modern Russian society. "Everybody knows that there's a wave or epidemic of spy mania or spy frenzy sweeping over Russia," Pasko wrote, "and if you look at history you can find an explanation. Suddenly the FSB have started realizing that they have to show society that they are earning their keep, that their huge numbers [budgets] are justified." The phrase "spy mania" soon reentered the lexicon. The publication of the book was timed to put pressure on the appeals court, which was shortly due to hand down its decision. A few days after the publication date, that court ordered Pasko to stand trial once again in a lower military court for abuse of office, on the grounds that this particular charge against him had not been thoroughly enough examined during the first trial. "This is a death sentence," Pasko told the press upon learning that he would have to stand trial again. "Russia is turning into a torture chamber."

Another high-profile case running while mine was going on was Valentin Moiseyev's, held on the fourth floor of the Moscow City Court building. I occasionally saw him and exchanged a few words with him on the way to and from the courthouse. Many of the tactics used by the prosecution in this, Valentin's second trial, were even worse than those used in mine. For instance, he and his lawyers were not permitted to see the charges against him, because the indictment itself supposedly contained secret material. His lawyers were not even permitted to show him documents that his defense team had prepared for the trial, even after these documents had first been submitted for review to the Lefortovo administration. (And this, when the key reason for the overturn of his first trial's verdict was that the lower court had been too vague about what secrets Valentin had allegedly given away to a South Korean diplomat!)

Despite the efforts of the prosecutors, the FSB, and Putin, the judge in Valentin's trial gave out signs that the court might not convict him this time around. Few cynics were surprised when, shortly before the verdict was to be rendered, there was a public announcement that the judge in the Moiseyev trial had fallen ill, and the conduct of the trial had been turned over to a second judge — who would have to take much more time to consider its myriad elements. During that review, of course, Valentin Moiseyev would remain in Lefortovo.

None of these cases was good news for me. The prosecutors' failures to win easy convictions in three high-profile cases — Moiseyev, Pasko, and the allied case of Alexander Nikitin — meant there was even more pressure on them to make certain that nothing would stand in the way of convicting me. I also feared that the tactics that they had recently used to prevent Pasko and Moiseyev from being freed after appeals might as readily be used against me.

One night, the popular *Kukli* puppets television program featured a skit involving me. Puppets representing Putin and Clinton are having a summit meeting — as the actual leaders had just finished doing in Brunei — and reach an impasse. Puppet Putin hands puppet Clinton a letter from Edmond Pope, which Clinton reads aloud: "Considering all the difficulties with the American presidential election, I've decided after all to remain in Russia." The Clinton puppet falls to the ground in a swoon. End of program.

Pavel and Andrey learned from their sources that Plotnikov's son had been a member of the FSB team that had first investigated my case. They didn't know precisely what role the son had played, but that did not matter, Pavel told me, because the law was clear: no officer of the court was permitted to take part in a trial if a family member was involved in the case in any way. In Judge Nina's courtroom, Pavel made a motion to remove Oleg Plotnikov as prosecutor in my case because of conflict of interest — a motion that if granted could result in an immediate end to this trial and the need to retry the case in its entirety, since Plotnikov's bias would be understood as having tainted his questioning of every witness.

This motion was a tremendous threat to the prosecutor's office, the FSB, and the judges. In immediate response, Comrade Barkova suspended the trial so that she would not have to rule on the motion. That was on a Friday. On Monday, the next court session, a new prosecutor showed up. His name was Yuri Volgin, and he reported casually that over the weekend, Comrade Plotnikov had suffered a heart attack and was now in the hospital, and that he, Volgin, would replace Plotnikov at the trial. The defense motion to remove Plotnikov thus became moot. Nonetheless, Andrey made a motion that the entire proceedings to this point be vacated due to Plotnikov's obvious contravention of the law; Comrade Nina wasted no time in rejecting this motion. Pavel reacted quickly, going on NTV that evening to challenge Volgin to admit or refute that Plotnikov had been replaced because of the family link. Volgin refused all invitations to share airtime with Astakhov. But Volgin did tell reporters, "I am not checking on family relationships, and any comments can be made by me only during the trial."

As we watched and listened to Volgin in the courtroom over the course of the next few days, it became apparent to all of us, including Judge Nina, that he was second-string, and that his ineptness would be on constant display throughout the rest of the trial. Shortly after Volgin took over, Barkova seized and held the reins of the proceedings, leaving very little for the prosecution to do. She spent most of the court's days reading aloud through the ten volumes of protocols.

I could see no reason for this procedure except to take up time. It reinforced my belief that the end of the trial was programmed to coincide with the settling of the American presidential election. Since that election was still in doubt in the middle of November, Comrade Judge Nina would delay matters by reading to the kindergarten. But almost as soon as she had begun her recitation, she had occasion to quote from an early protocol, and in the course of it she mentioned the name Plotnikov as one of the investigators, thus inadvertently confirming what we had alleged in our motion for the removal of the prosecutor. I laughed out loud at hearing this name, and so did my lawyers, but Volgin acted as though he was the monkey that heard no evil, and Judge Nina went right on reading as though nothing had happened.

Andrey introduced a new motion, demanding that Judge Barkova be removed from the trial because of the flagrant bias of her wholesale refusals of the many motions made by the defense. Nina stared at Andrey with her mouth wide open when he offered this motion, but it had to be considered, so she let her compatriots on the bench chew it over for a while. The older man and older woman retreated to chambers, while she stayed in her chair and glared at Andrey; when his cell phone beeped, she barked at him and pounded her fist on her desk. Shortly, the two bookend judges returned and announced their decision that she ought to remain as the chief judge for the trial. After this, Barkova's attitude toward Andrey hardened even further.

Pavel, Andrey, Keith McClellan, and Peterson's office had worked very hard to obtain statements from Penn State, ARL, and Dan Kiely that would be of assistance to my defense. After much negotiation, and very late in the game, Kiely put his name to a letter to the court in which he recanted the information in his protocol, saying, much as Babkin had, that it had been obtained from him under duress. In Dan's instance, he might also legitimately claim that he had not known what he was signing, because he could not read the protocol document. Kiely did not go so far as to write, as we had hoped he would, that many of the materials in his briefcase and computer were matters that Ed Pope knew nothing about, but at least he disavowed the protocol that the prosecution said implicated both him and me in espionage.

On November 23, Thanksgiving Day, the court session was short but fruitful. Unexpectedly, Judge Barkova agreed with a defense motion to exclude the Kiely protocols. This meant that the statement containing the only American admissions against me was thrown out of the trial. It ought to have been a triumph for me, but it was merely a small bump of joy along the road to a certain conviction, because while she excluded the protocol, the judge refused to throw out the materials that had been seized from Kiely.

Another hollow triumph occurred when Judge Barkova accepted into the court records another item we had long sought: a document from the Bauman review committee, setting forth its findings that there was no classified material in the Babkin reports and authorizing the sending of those reports out of the country. Here again, this document, which would have blown open the case in an American court because it upheld our contention that there had been nothing secret in the Babkin papers, was simply accepted into the record by Comrade Barkova, its legal implications ignored.

That evening, in the cell, I received the Thanksgiving dinners that the American embassy had prepared for me and my three cell mates. The dinners came in three boxes each and included as many American favorites as could be found. The amount of food would have kept us in culinary contentment for several days. But the guards insisted we eat all of it on Thanksgiving evening, so we took our time and stuffed ourselves, managing to secrete just a bit of the food around the cell for later consumption. During our lengthy dinner, my companions ribbed me a bit about the American election snafu, which many Russians took as evidence that the U.S. was almost as corrupt and mixed-up a society as Russia. One translated for me a joke from a Russian-language newspaper: Putin sends an observer to Chicago to watch the American presidential election, and three days after the votes are cast the observer reports back to the Kremlin that Putin is in the lead.

My Thanksgiving holiday was brief, because the next morning, in court, the prosecutor unexpectedly introduced the fact that there was to be a civil suit against me, brought by the economics section of the FSB on behalf of the Russian navy, for seven billion rubles (or $250 million dollars), the amount of money they said had been stolen from the Russian navy by my alleged purloining of the secrets of the Shkval. The monetary figure represented the development, testing, and manufacturing costs of the underwater missile.

This was so outrageous a charge that even Comrade Barkova decided that it could not be included in the proceedings of this court, on the grounds that it had not been introduced early enough in the trial. Since the charge would eventually be brought in a civil court, she did not have the power to dismiss it, but she opined that it would soon be brought to fruition in the proper venue. For emphasis, she gave me her best evil-eye glare.

That evening, brooding over the $250 million civil suit, my anger finally got the better of me, and I took out my frustration by kicking the seat of the *parasha,* the bucket-shaped appliance that served as a toilet in our cell, so hard that it broke. My cell mates applauded and suggested that the prosecutors file an additional $250 million suit for this crime. We all had a good laugh and then they helped me fix the seat so that it did not appear to have been assaulted. They also made a sign, in Russian, that celebrated me as a *parasha*-basher, and with a heavy black pen I autographed the undersides of the seat. We agreed that though I might be innocent of the charges brought against me in court, I was definitely guilty of defiling Russian property — in the form of a wooden toilet seat.

I confided to my diary that this spurious $250 million charge had finally pushed me over the edge. I now hated the bastards for the agony and suffering they were continuing to cause to my family, and equally for the damage they were doing to bilateral relations between the U.S. and Russia. I could no longer avoid the conclusion, as I put it in my diary, that the FSB and its allies were "actively trying to drive a wedge between our two countries and peoples," a wedge that would produce suffering for millions of people in Russia should they be deprived of American aid and American business development. For that crime — and not for what they had done to me personally — I hoped the FSB and its political and criminal allies would all burn in hell.

It was time for closing statements. The prosecution's final words were read by Yuri Volgin, but had obviously been prepared by the FSB, since Yuri tripped over some of the longer words that were unfamiliar to him. Some of what he had to recite was so propagandistic that it made Andrey and Pavel chortle. "The U.S. claims to be a partner of ours, yet they send spies like Pope into our midst." "The FSB investigators had discovered in Edmond Pope a Trojan horse of the American secret services." Volgin's claim that I had a close personal relationship with Babkin even made the judges giggle, since anyone could see from the hidden-recorder videotape that Babkin and I could only communicate through an interpreter and that I was keeping the old professor at arm's length. Another laugh-out-loud Volgin/FSB claim was that my espionage had continued up to and during the trial itself, making itself felt in the questions I asked of Shakidzhanov on the stand about whether he had the authority to sell Shkvals abroad, and about a fire-fighting system. The FSB evidently considered me to be a super spy able to commit espionage within the confines of a closed courtroom! Such wretched excess was dismissable, but I was very disturbed by Volgin's insistence that the court find me guilty and sentence me to the maximum, twenty years at hard labor, and fine me $250 million. Every indication that we had had prior to this time suggested that there would be some lesser sentence. I began to worry again that I would never leave Russia alive.

Now it was the defense's turn. Andrey Andrusenko started, Pavel would go second, and I would be allowed the last word for our side. On December 1, Andrey came into court with an astonishing document as an exhibit tied to his closing statement. It was a loose-leaf book that he had prepared in which the five Babkin reports were printed, and alongside them, the excerpts from open-source materials on which Babkin and Myandin had based every bit of their work. This book was even a surprise to me, as Andrey had kept its construction secret. But it was a stunning, provocative document, one that in any other court might have sunk the prosecution's case. Here it provoked an immediate recess, during which Alyosha ran out of the room, probably to consult higher authorities. He soon returned and went into the judge's chambers together with Volgin.

True to form and obedient to FSB authority, as soon as the trial resumed Judge Nina Barkova refused to permit Andrey to place the loose-leaf book into evidence. But Andrey had another card to play, a letter that the prosecution had already introduced as an exhibit; it was from Russia's foreign-intelligence service, and it said that Shkvals were in the possession of other countries, such as Ukraine, Kyrgyzstan, and Kazakhstan, and that the last-named had sold fifty of its Shkvals to China.* Therefore, Andrey said, the Russian Federation could not continue to claim that the designs for the Shkval were a state secret. Andrey also forcefully reiterated the theme that this trial had been a witch-hunt, not a fact-finding expedition, and that the FSB's investigation had been a similarly flawed attempt to fix blame,

*Newspaper reports give this number as forty, but the document shown to me by the FSB had it as fifty.

disguised as an inquiry into what had happened. The facts proved, Andrey said, that no crime had been committed or contemplated. The security of the Russian Federation had never been breached, and Ed Pope had not even attempted to breach it.

Pavel now took the stage. He had come to court for his closing statement dressed as nattily as I'd ever seen him, and he put on a performance that surprised more than a few people inside and outside the courtroom. He announced that he had composed an ode in the Shakespearean mode concerning me, the trial, the issues, the judges, et cetera, and proceeded to declaim it from his notes. From the reactions of Russian speakers, this ode featured humor and a goodly portion of irreverence, and was a carefully calculated slap in the face of the judges, particularly Nina Barkova. Nevertheless, his recitation provoked the only real smile I ever saw on her face. She even complimented Pavel on his use of iambic pentameter.

After a few moments of trying to translate Pavel's poetry into English, Alyosha threw up his hands, told me he couldn't do it properly, and thereafter kept quiet. Pavel, when he finished speaking, would not tell me precisely what was in the ode, and later released to the press only the beginning of it:

> *I call on you to open your eyes, tune in your ears,*
> *and speak the truth from your lips.*
> *There is one truth: he is not guilty..*
> *To acquit him is a societal task, not just for you*
> *and me, but for all Russian society.*

He told me he was planning to print the rest of the ode in a book that he was writing, which would contain the stories of his work to defend me and Vladimir Gusinsky.

All that remained of the trial procedure was for me to give a closing statement and for the court to render its verdict; both would occur after a weekend recess.

During the weekend, there were several developments of interest. The Putin government chose this moment to unilaterally terminate the Gore-Chernomyrdin agreement of 1995 that had committed the Russian Federation to stop selling arms to Iran. Western news services also began to run stories, based on interviews with business sources in Moscow, stressing that my conviction would deter American businesses from doing further commerce in Russia. Also over that weekend, Cheri, John Peterson, and his associate Jen Bennett arrived in Moscow to be with me at the close of the trial.

Cheri and I had a brief visit in prison — only the fourth in the eight months that I had been incarcerated in Russia. We were both very tense, preoccupied with the hope that all this would soon be over, one way or another. Cheri later said that she was shocked by my physical and mental condition, that I had been unable to speak coherently or to write notes. John Peterson spoke for us all when he told the press, "Edmond Pope can't stand much more. It's been 244 days of being deprived of life." Cheri informed reporters that if there was a guilty verdict and no subsequent action to send me home for humanitarian reasons, she would take a short while to collect her thoughts and would then move to Russia to be near where I was next imprisoned. As she took part in a radio call-in show, an elderly World War II veteran told Cheri that he would offer to serve my prison sentence so that I could go home to my family.

On Wednesday, December 6, it was my turn to speak. The court would not allow Cheri or John Peterson or June Kunsman to come into the courtroom, so Cheri stood outside in the hallway, where she could see me at a distance before the doors were closed.

I used my privilege as the defendant to introduce into evidence Andrey's 300-page loose-leaf book — the judge did not have the right to refuse to admit it if it came from me, but she tried: when Andrey first attempted to tell me to introduce it, she ordered him not to speak to me; when he wrote out the instruction on a note and held it toward me, she ordered the note seized. But I understood what needed to be done, made my submission, and so there entered the official record a lengthy document detailing whence every fact in the supposedly incriminating Babkin papers had been drawn, and showing that his sources were all unclassified.

I then went on to present the statement I had worked on for some time. It was based mostly on the testimony of Myandin, Babkin, Pavlikhin, and others who had confirmed my intentions, actions, and the unclassified nature of what I had sought and what Babkin had delivered. Seven different people directly involved in the preparation or vetting of the Babkin reports, I said, had testified that they contained no classified material. Moreover, the "expert" commission had never been asked to verify whether the information in the Babkin papers was based on open-source material. I trumpeted that the Kiely protocols had been thrown out of the case by the judges. I lamented that Bolshov had not been permitted to testify. I queried the interpretation of the hidden-recorder videotape from the Sayani hotel. I tweaked Volgin on his refusal to come clean about Plotnikov's family connections to my case.

In conclusion, I said that I stood falsely accused of espionage, but was more concerned with the larger issue: that those in Russia who needed to accuse a foreigner of espionage were causing great damage to the Russian Federation by taking away from its people great possibilities of employment and the economic and social benefit to be gained from scientific exchange with the research institutes and entrepreneurial companies of the West. "I urge you to turn your attention to meaningful and needed reform rather than to continue to blame outsiders and individual Russian citizens for problems that are systemic in nature." Finally, I told the court that I had been in jail for eight months "for a crime that was neither committed nor contemplated," and I asked the judges to acquit and release me because that was the decision that was "best for both Russia and America."

* * *

Two hours after my closing statement, Judge Nina Barkova sent word that she was about to reconvene the trial to issue the verdict in Case #8 of the year 2000 for the Moscow City Court. I was then in Lefortovo, on a lunch break.

As I came out of my cell, it was to a glaring light illuminating the scene for a television camera — just one, from RTR, which had been "chosen" by the other broadcasters to act as a pool camera whose feed could be used by all. More massed press awaited in the hallways of the Moscow City Court building. And for the first time since the start of the trial, the courtroom was packed to capacity, about fifty people including Cheri, John Peterson, Jen Bennett, Bob Ferguson, June Kunsman from the embassy, reporters, and onlookers, though no cameras were allowed. Cheri elbowed her way near to the defendant's cage, and we held hands through the bars. The guard tried to prevent her from doing this, but she glared fiercely at him and so did I, and eventually he backed down, leaving us together for the moment.

The verdict was contained in an eighteen-page document that in the past two hours Judge Nina had ostensibly written, but also had translated into English, supposedly after she and her colleagues had digested — among other things — the 300-page book I had that morning put into evidence. The printed verdict's very existence so soon after the closing argument shouted that it must have been prepared well in advance. Pavel would later comment to reporters that in this, the most important espionage trial in forty years, everything had been rushed: the investigation, the trial process, and most of all, the writing of the verdict, which he needled had been done in "world's record" time.

The verdict largely replicated the language and conclusions of the indictment. It was no surprise to me or to anyone else that the court found me "guilty in committing espionage" during the period between October 1996 and July 20, 1999, actions that had the effect of "handing over to foreign organizations, collected technical documentation comprising state secret information to use the information to the prejudice of external security of the Russian Federation." The "foreign organizations" were specified as "military-industrial establishment enterprises," presumably those of the United States, although that was not stated explicitly, and the "state secret information" concerned the development of the M-5 Shkval, its design and update, and its solid missile fuels. The "technical documentation" through which I had supposedly obtained this secret information were the five Babkin reports.

The testimony of various witnesses before the court was cited in support of the conclusions, but the verdict's characterization of that testimony was corrupt. For instance, it twisted what had been said by Pavlikhin, omitting entirely any mention of his testimony about the Bauman review committee, and merely reporting that Pavlikhin had signed the Babkin papers. Although Babkin had recanted his early testimony, both by letter and by his later testimony in court, the verdict writers said they would accept the conclusions of the Babkin protocol and reject his in-court testimony. "The court explains the evidence, given at the court session by Babkin who wanted to diminish the degree and the extent of the charge brought against Pope, by the fact that now criminal proceedings are instituted against Babkin himself [under] Article 275 of the RF Criminal Code."

The verdict concluded that my supposed espionage would enable "a considerable shortening of terms, reduction of material and human costs for foreign specialists in developing missiles analogous to [Shkval] and its modifications," and that "arming the national Navy of foreign states with new weapons on the basis of the obtained information . . . will lower the combat stability of the Russian Navy." Cited as evidence of my supposed espionage activities were my expired service identification card, my Navy Federal Credit Union ATM card, and my membership card in the American Legion.

Then came the section on "punishment."

As we waited for this in court, Cheri and I holding hands tightly through the wires of the defendant's cage, I expected the minimum sentence, ten years. What came through, in translation, was weird:

because the court found the crime a grave one that represented "serious social danger," it had taken my "personality" into account and determined that the awards that I had received from the U.S. Navy constituted "received knowledge and experience" that I had used "to the detriment of the Russian Federation." Therefore, I was sentenced to the maximum penalty provided by the law, twenty years at hard labor, with specifically reduced opportunities for contact with my family.

Despite my prior certainty that I would be convicted, the maximum sentence stunned me, and it stunned Cheri. For the moment, we could do nothing but clutch hands more tightly and look at each other in complete disbelief.

The guards ripped us apart, sundering that brief embrace, and hustled me out of the courtroom, down the back stairways, and out into the bread truck to return me to Lefortovo.

CHAPTER 11

WAITING FOR PUTIN

As the bread truck pulled up inside Lefortovo, I was met by a welcoming committee of television lights and a camera. I recognized a young female television journalist from RTR, who wanted an interview. She was polite and friendly, and asked the obvious questions, such as "How do you feel?" and "What do you think of the verdict?" I felt like no more than a confused bundle of nerves, but certainly didn't want to say so, or even to answer her questions. I gave a few peremptory responses, then told the guards I wanted to return to my cell. They led me away, but not to the cell. Instead, they parked me in a shabby lounge and seated me in a low and uncomfortable lounge chair. Little Miss KGB appeared again with her crew. I decided to answer some of her questions, in the belief that this would aid me in my request for a pardon (and the belief that not answering might hinder a pardon) and in the hope that after answering, I would be left alone. In her introduction, though, she addressed me as "America's James Bond," and I had to insist, for the umpteenth time, that I was no James Bond.

I told her that I regretted that this "incident" — the entire case — had occurred; said that it had damaged relations between our countries but that we should put it behind us and continue to build a partnership between the Russian Federation and the United States of America; and reiterated that I had respect and admiration for the Russian people. She surprised me by a hint that I would be released soon, and asked if I would then consider returning to Russia. I told her I would not. I declined her offer to say something to whomever would be the next American president, or to President Putin, since I thought it inappropriate to do so, but said that, yes, I would pray to God for a solution to the situation in which I found myself. When the interview was finished, the woman who had begun it by accusing me of being a James Bond gave me a hug and told me that she wished the best for me. Her behavior was a perfect reflection of the two-sided nature of everything in Russia.

I spent a fitful, sleepless night, depressed about the court's sentence and wondering if it might prevent Putin from freeing me anytime soon. I still felt very ill, aches and pains everywhere, ominous lesions on my face and neck, and was worried about what further incarceration might do to my health. Suppose Putin decided that the length of the sentence meant he ought to wait another year before releasing me?

My daily newspaper arrived the next morning. In addition to a report on the trial verdict, the *Moscow Times* printed an editorial entitled "Pope Verdict Caps Shabby, Unfair Trial." It skewered the verdict and the sentence as "the inevitable outcome of a judicial process that was nothing short of a travesty. We do not know whether Pope is guilty or not of violating any laws, but we are certain that his guilt was not established during his trial." The editorial went on to enumerate the many instances of the judge's bias, her denial of defense motions, the basing of the charges on secret documents, the importance of the Babkin recantation and of the fact that Babkin's reports were based on open-source material, et cetera, and summed up the trial as a "sham, reminiscent of Stalin-era legal proceedings in which the state stopped at nothing to get the result that it sought." Accordingly, the newspaper asserted, the verdict should be overturned.

Later, I would learn that many American newspapers carried front-page articles about my trial and conviction, and some also had editorials criticizing the trial and demanding my release. Liliya Shevtsova, an analyst at the Carnegie Institute's Moscow branch, was quoted as saying the verdict was part of the Kremlin's drive to restore national pride through emblems, of a piece with the recent decision to again officially use the old Soviet national anthem. The National Security Council summed up the official feeling of the administration in a statement that "the verdict is unjustifiable, it is flat-out wrong, and it has cast a shadow over our relationship."

Now public officials in the U.S. who had initially refused to help us, or had been reluctant to lead the fight, such as Senators Arlen Specter and Rick Santorum of Pennsylvania, started to jump on the bandwagon, calling for an immediate pardon and saying bad things about Russian justice and President Putin. In Moscow, on television and radio there were hourly news bulletins about Secretary Albright's phone calls to Foreign Minister Ivanov, the congressional threat to terminate U.S. aid to Russia, and the deteriorating state of my health.

While mulling all this, I was summoned to the offices of the warden, Colonel Kuryushin. A handsome, almost elegant man in a well-cut suit and with a courteous manner, Kuryushin was the sort of person that under other circumstances I would want to befriend. He was very pleasant, although he did not shake hands, and through an interpreter he explained that I was a problem to him, an administrative problem, so he would like to see me released from his prison, for humanitarian reasons, as soon as possible. He wanted me to write a letter to President Putin requesting a pardon. I had already written such a letter, which said: "I request that I be released from prison to return to my family in Pennsylvania and receive health care. I am not well. I need immediate medical care." Patiently, the colonel told me to write a new draft, including the word "pardon," and certain other key phrases, so that it read:

> I request that you give me a pardon and release me from prison so I may return to Pennsylvania to my family. I do not feel well and I need urgent medical care with my regular doctor and specialists. I beg you to solve this question as soon as possible

because my father is terminally ill and I want to see him for the last time.

I agreed to do as he suggested. He gave the new letter to an aide for immediate delivery to the president, and then insisted that I write a second letter, to Judge Barkova's court, in which I was to assert that in consideration of the possibility of a pardon from Putin, I would not seek an appeal of the Moscow City Court's verdict. Kuryushin told me this second letter would force Barkova to turn over control of me to the prison administration, which would make visits by Cheri easier, and it would put more pressure on Putin to immediately accede to my request for a pardon. I wrote the letter but declined to send it, on the grounds that I needed to confer with my family, lawyers, John Peterson, and the American embassy before renouncing an appeal. He agreed to call Barkova for permission to phone the embassy on my behalf to get this ball rolling.

In the afternoon, I was summoned from my cell again to meet Andrey Andrusenko. He had been waiting since morning, he said, to see me. I began to become paranoid, thinking now that Kuryushin had set me up, trying to force me to write the letters before he would let my lawyer advise me about them. But Andrey assured me that the no-appeal letter was a good idea, and I agreed to have it sent. He also told me that Judge Nina had refused a request by Cheri to visit me today — information that upheld Kuryushin's reason to have a letter that would transfer jurisdiction over me from the court to the prison, making it easier for Cheri to see me.

That evening, Russian television news reported that President Putin had received my request and was turning it over to a special pardon commission. I could hardly contain my excitement, but felt I could not yet rejoice wholeheartedly, not until a pardon had been recommended and Putin had agreed to it.

Cheri, John Peterson, and officials at the American embassy had received similar assurances about the pardon, but neither they nor the people who were giving them this information properly assessed the potential for intransigence and resistance to the president's wishes within the FSB and the Moscow City Court.

The Pardon Commission had been established by Gorbachev during the last period of the Soviet Union, when the country was crumbling, and Yeltsin reinvigorated it when he replaced Gorbachev in 1991. Anatoly Pristavkin was its chairman. Pristavkin, author of a best-selling novel, *A Golden Cloud There Rested,* recruited other prominent intellectuals and human rights activists as volunteers. The group was composed of seventeen members, including two more well-known figures, Marietta Chudakova, the biographer of Mikhail Bulgakov, and theater director Mark Rakovsky. Other members were a former judge, a surgeon, a philosopher, and a Russian Orthodox priest. More than 5,000 applications arrived in the commission's offices each week, and a staff culled 250 of these for serious consideration by the commission at each meeting. Every Tuesday the commission members met and decided on the fate of those applicants. Pristavkin had been quoted as asserting that the Russian police, the FSB, the prosecutors, and the courts usually let the worst criminals off without jail time but pursued and jailed small criminals mercilessly, because "that's how they want to convince the public they are fighting crime." Accordingly, the commission would pardon many of those whom they believed had been unfairly punished.

The commission did not always have smooth sailing. For two years during Yeltsin's time in office, 1994–1996, the flood of pardon requests dwindled to a trickle, because chief state prosecutor Yury Skuratov wanted convicts to stay in prison. Pristavkin eventually succeeded in reopening the flood gates by personally interceding with Yeltsin's administrative chief. When Putin took over from Yeltsin, there was some doubt as to whether he would continue the Yeltsin tradition of granting pardons that the commission recommended, but in Putin's first year in office he did, giving amnesty to 12,500 convicted prisoners — virtually all of those whose pardons had been recommended by the commission.

The Pardon Commission generally met on Tuesdays, but in order to consider my case, it assembled on a Saturday, December 9 — one day after they received my pardon request from Putin — an action that underlined the government's intent to have no further delay in addressing the matter of Edmond D. Pope.

The decision reached by the commission about me that Saturday morning, Pristavkin would later tell reporters, was unanimous: they recommended that Putin pardon me, and do so quickly. While Pristavkin stressed to reporters, "We did not judge the ruling of the court. We made our conclusions on humanitarian grounds," Maria Chudakova went much further. She told reporters that the trial demonstrated that "the investigative organs of our country still bear the marks of the Soviet system, more so than society in general," and that the televised parts of the proceedings demonstrated "spy mania, which we all know about." Theater director Rakovsky echoed Chudakova, adding that "Pope is not Gary Powers," and recommending that Putin show clemency to me as a way of demonstrating that Russia had stepped away from the Cold War mentality.

Around midday, the news of the Pardon Commission's decision about me began hitting the airwaves. The recommendation was unanimous, and the commission's letter to Putin said that I should be pardoned and released immediately. I breathed a huge sigh of relief. My cell mates were elated. They hugged me, told me how glad they were that I'd be going home for Christmas.

Of course, Putin had not yet announced that he would grant the pardon, but we all believed that to be a formality. Now all I could think of was how soon I would be able to leave this awful place, step aboard a plane, and get out of Russia. If anything, my feeling of desperation increased once I knew that I would get out but was waiting for my release. That feeling intensified during the visit that afternoon of Andrey and Pavel.

There was a snag, they told me. It would take seven days for the paperwork to be properly done and for me to be released. The target date was the 15th — but Cheri's visa expired on the 14th, and given the past behavior of the Ministry of Foreign Affairs and the FSB, it was unlikely that her visa would be extended. My mind flashed to the grim joke about Russian mathematics: Question: How much is two plus two? Response: Whatever the Comrade wants the answer to be.

But this exhausting Saturday was still not over. Around ten in the evening — a very strange moment, since bedtime was about to be announced — I was summoned again to the office of Colonel Kuryushin. He informed me that President Clinton had just spoken by phone with President Putin, and Secretary Albright with Foreign Minister Ivanov, and they all wanted to know if I desired or required immediate hospitalization.

Go to a hospital in Moscow, a setting in which many people have been known to die suddenly or disappear or become so crazy that they then have to be taken to mental institutions? No, thank you, Comrade!

At Kuryushin's request, I wrote out a declaration that I did not require immediate hospitalization in Russia but that I did need to be released as soon as possible so that I could go to an English-speaking hospital chosen by my wife, Cheri. This letter in hand, Kuryushin picked up the phone and spoke directly to President Putin, telling him (I presumed) what I had agreed to. Something would happen tomorrow, I was told as I was returned to my cell for another night.

On Sunday, President Putin announced that he was acceding to the recommendation of the Pardon Commission and granting me a pardon.

I felt an immense sense of relief. At last I would be going home. But it didn't happen Monday morning, and I started to worry again. Why, if the pardon had been granted, could I not immediately leave the prison? As always, there was a reason — not a good one, but enough of one so that the FSB and the courts could buffet me under the guise of pretending to abide by the rules. It seemed that a seven-day window during which I could have filed an appeal had to elapse before a pardon could be granted and I could be released. The FSB contended that the seven-day window had begun from the latest possible moment, when I wrote a letter declining to take the appeal. That would mean, of course, that Cheri's visa would expire before I could be released, a petty determination that even flew in the face of what Putin himself seemed to want. Pavel went into court to argue that the seven days had begun earlier, when the verdict was translated for me in the trial courtroom, because I had told him then that I wanted a pardon, not an appeal.

There was enough of a delay so that John Peterson had to return to the U.S. to take part in congressional work, only to have to deal with another problem: the State Department and its counterpart, the Foreign Ministry, wanted to get me out of Russia, when the moment came, by putting me on a commercial Aeroflot flight. Peterson's staff and my family objected to this for a variety of reasons, and negotiations about just how to take me out of Russia went on almost until the last minute.

Jen Bennett and Cheri spent many hours in a room in the Kempinski — a luxurious prison, as it were, since they could not go out of the hotel without being subjected to media coverage and to shadowing by the FSB. During this period, Jen set up interviews for Cheri and Congressman Peterson with both *Good Morning America* on ABC and *Today* on NBC, minutes apart. When the entourage was traveling by limousines between one studio and the next, they were stopped by FSB officers. There seemed no reason for the halt except to harass the Americans; NBC's Moscow producer, who was in a car leading the caravan, stepped out and spoke sharply to the officers, convincing them to let the cars and their important occupants proceed.

As the group was walking away from the studios, a Russian woman came up to them on the street and asked Cheri if she would accompany her to a church. Cheri wanted very much to go to a church to pray and told the woman that she would do so, but not while they were being followed everywhere by cameras and reporters. She hoped to go to a church perhaps in the early evening, or at a time when it might be mostly deserted. Eventually, though, she gave up this idea, as Jen Bennett thought it would be impossible to prevent the press from finding out where Cheri was going and turning the church visit into a photo opportunity.

During this tense period, my wife told an interviewer, "Sometimes I feel like there's not even a square inch of Cheri left. I don't feel like I've been able to be the private me, to have something that's not looked at — and that's very, very difficult for me."

The American election was still not officially settled, though a series of decisions in the Florida and the U.S. Supreme Courts made it more and more likely that Governor Bush of Texas would be declared the winner by the electoral college. That eventuality was scheduled for December 12, a date that was nearing. It now seemed most likely that the precise timing of my release from Russia was to coincide with the official certification of Bush as president.

Peterson received a call from his fellow Pennsylvania congressman, Curt Weldon. Weldon informed Peterson that the Republican congressional leadership had arranged to have a military plane to fly me and Cheri home to the U.S. from Moscow, and had also arranged for Peterson to walk us out on the tarmac, but that we would be met at the other end and welcomed home by Weldon and the congressional leadership. Peterson informed Weldon that a plane had already been arranged by the military — but not a military aircraft, since that would give the impression that I had indeed been working for the DIA or the CIA. That rented executive jet was waiting in Germany for instructions, Peterson went on, and after picking me up would take me to a military hospital in Germany for medical assistance, tests, and debriefings. It would be difficult and embarrassing to cancel these arrangements, Peterson summed up. Weldon reluctantly agreed that Peterson's arrangements could stay in place and the other plane would be canceled.

Meanwhile, Putin left for Cuba, to visit that great Communist Fidel Castro, and would remain there, on public display, during the moment of my release — perhaps so he would not be near any Western newsmen who might ask him a question about Ed Pope.

Late on December 13, I was told by the prison administrators that they were going to transfer me to a single-person cell, near the storage lockers, where I could sort and pack my belongings in the expectation that I would be released on the following day. Hurriedly, I distributed all of my cached food and most of my extra clothes to my cell mates. They went wild over the clothes: "Look, this shirt was once worn by Ed Pope!" We four embraced and pledged to meet again at a better place and under better circumstances.

At eight in the evening I was put in a cell on the first floor, and doors were opened so that I could go between it and the next cell, into which all my things from storage had been delivered. I had promised my television set to the American embassy for donation to a charitable institution. No more than fifteen minutes had gone by when the cell door opened, two guards walked in, and delivered back to me all the clothes that I had tried to distribute to my former cell mates. Not permitted! That broke my heart: what cruel bastards these FSB men were, not to allow prisoners — detainees who had not yet been convicted in courts of law — to have the clothing that I had freely and happily given to them. I went through my accumulated things, such as the dozens of letters that had been sent to me but which I had not been permitted to keep in the cell. There were some that had never been delivered to me, though they had been transmitted to the prison months ago. Books. Clothes. The cards sent to me by a school class in Minnesota. The other, small TV that an early interrogator had given me. The suitcases that I had long ago stashed at the Sayani Hotel. Even the more than 600 pages of notebooks that I had compiled while in prison, which contained the only available record of my interrogations and trial. Among the treasures I found was a miniature bottle of vodka, which I proceeded to down, the first alcohol I had drunk in more than eight months. I slept very soundly.

While I was sleeping off my vodka and nervous tension, late in the evening Jen Bennett received a call that brought her to tears: the Foreign Ministry, in cahoots with the FSB and the court, had decided that the seven days had not elapsed and so the government was not going to release me from Lefortovo tomorrow morning (Thursday), as planned; rather, they would keep me until the following Monday.

Bennett cried but did not panic. She called June Kunsman of the embassy and complained bitterly: Peterson and the Pope family, Jen said, had played the game precisely as directed to by the State Department, to minimize embarrassment to the Russian Federation, and now the Russians were being just as obstructionist as if Peterson and the Popes had been uncooperative.

"You're right," Kunsman said, and started to work. She galvanized Ambassador Collins, and together they called the Ministry of Foreign Affairs and insisted on an immediate meeting. Near midnight, the two groups met and worked out a deal, whose particulars were relayed by Kunsman to Bennett in the early hours of the morning: I would be released from prison that morning, but in return, Bennett had to agree not to alert the press corps about the timing of the release. "I felt really bad about that," Bennett recalls, "because the press had been so uniformly helpful to us and sympathetic to Ed. But it was the only way the Russians would let him out on Thursday, so we had to agree."

Later that morning, ignorant of the machinations of the night before, excited and oblivious to my surroundings, I was taken up again to Kuryushin's office, where there were several functionaries, a photographer, and an interpreter. On an earlier visit, the warden said that reporters and cameras might be waiting for my release and had asked if I wanted a press conference. No, I had told him. But he had seemed to want one, and I had agreed to prepare for him a written statement that he could use for such a conference. I handed it to him now.

An "open letter to the Russian people," it said that I had been "deeply troubled by the proceedings and the verdict handed down," and that among the reasons for declining to take an appeal to the Supreme Court was that "I have little confidence that I would receive a fair hearing anywhere within the present Russian judicial system." Rather than holding the Russian people as individuals responsible for "the abuses of justice and human rights that were committed against me," I had been heartened by encounters with many compassionate individuals, and by the comments of the Pardon Commission. I linked the injustice I had endured with the injustice more regularly "suffered by the Russian people under this system," and urged them to use my high-profile case "to demand and accelerate badly needed reform in your criminal justice system." The virtues and talented attributes of individual Russians would never be fully realized "as long as your system continues to repress and unduly persecute its citizens and deny them basic human rights." The Russians who had been accused along with me, I asserted, were innocent of the charges, and should "not [be] prosecuted or otherwise abused for problems that are basically systemic in nature." I urged the entire Russian Federation to put this case behind them, and to "continue your difficult task in transition toward a democratic society," one that recognized that its people must be given the "respect and support" they deserved in order for Russia to "realize her full and vast potential."

When Kuryushin heard the Russian translation of this statement, he simply nodded, and I understood that my words would not then see the light of day inside Russia: the statement contained too much criticism of the system for Kuryushin to risk releasing it.

But he did read out for me the pardon signed by Putin and handed me the Russian and English copies of the verdict in my case, which I was permitted to keep. My mind accelerated into the blue; the ordeal was nearly over. I listened without really hearing as Kuryushin told me that I had been a pain in the butt, but that he liked me and under other circumstances would have wanted to be my friend. He praised the courtesy I had shown to the prison administrators and said that the guards spoke highly of me. In closing, he invited me to come back to Russia and visit him, any time. I smiled, shook his hand, and was escorted out.

Then it was down the harshly lit corridors, through several electronic doors and barred gates, and out into a sunny, wintry day. The cold air felt good. A white American embassy Chevy Suburban sat inside the Lefortovo yard, and I could see Cheri and some other people in it, waiting for me. There was a television camera crew recording the scene — the same Little Miss KGB, with her saccharine smile — but everything else was a blur as I headed for the car. Someone told me they were going to take us in a caravan out a back exit, to avoid the reporters and television crews camped out at the main gate. I hoped Cheri would get out of the car and run to me, but they would not permit her to leave it while it was inside Lefortovo. I wondered who the FSB hated more, me or Cheri. Then I reached the white car, the door was opened, and Cheri and I embraced. I was almost home.

CHAPTER 12

Release, Recovery, Reconstruction

"Just tell me when we're out of Russian airspace," I requested of the pilot of the executive jet on which we were being whisked away from Moscow; I refused to eat or drink until that moment had been reached. Though weak, tired, and confused, I was giddy with excitement and relief, barely able to endure the questions of the American military psychologist and the probes of the medical doctor aboard the plane. They asked if the Russians had hurt me in any way, but what they really wanted to know was whether I had been tortured and divulged any secrets. "Nothing," I fairly shouted in triumph. "I told the bastards nothing."

After two hours, the officials let me go and sit with Cheri; I babbled to her, wanting to tell her everything that had happened to me, and peppered her with questions, aching to learn from her all that had gone on in the attempts to obtain my release. Who were the good guys? The bad guys? Who had done her wrong? The details that I began to learn in this conversation about her endeavors over the past eight months made me even more astonished and proud at what she had accomplished for me.

We had a laugh about the scene at Sheremetyevo airport, where Ambassador Collins had shown up to see us off, and so had Little

Miss KGB and her lone camera crew. We'd put up with them because both were just doing their jobs, but neither had done us any favors. The pilot sent back word through the stewardess that we were entering German airspace, and we had our celebratory popping of corks and downing of champagne. Not long afterward we touched down at the Ramstein Air Force Base, to find John Peterson waiting on the tarmac with a small military escort. "Welcome home," he said as he shook my hand and hugged me. I felt an immense sense of relief; for a man like me, who had spent so many years at U.S. military facilities in foreign countries, this air base in Germany felt indeed as safe as being back in Pennsylvania.

There was little additional fanfare, which was all right with me. By motorcade we traveled to the nearby Landstuhl hospital complex. For the next two days I went through a battery of medical tests, and two lesions were surgically removed from my face and neck; one proved on biopsy to be cancerous — though the "specialist" dermatologist in Moscow had sworn that it was not. My eyesight had deteriorated. So had my teeth, which would require a lot of dental work. The doctors determined that I was stable enough so that the more extensive tests on my hemangiopericytoma and my fragile immune system could wait until I reached the Bethesda Naval Hospital. Debriefings by the military were mercifully quick and perfunctory; more work on this, too, could wait until we were in the U.S. Cheri and I passed the nights in a suite so large and sumptuous that we joked that it must be used for honeymoons. Everyone at Landstuhl bent over backward to make us comfortable.

We enplaned again, on an American commercial carrier, for Oregon, accompanied until we reached Portland by two bodyguards. In Grants Pass my father was dying of multiple myeloma; transfusions and extra medications had been given to him to sustain him until I could return. He was very frail, in a wheelchair, only now and then fully aware of his surroundings. Sometimes, my mother said, he had been unable to recognize her or my sister, Brenda. But he knew who I was, and who Cheri was, and tears rolled down his cheeks as he exerted his remaining strength to squeeze our hands. It would be less than two months before he passed away; my mother and I were at his side then, too. After he was gone, I brought out of my pocket the little Bible that he and Mom had sent me in prison and looked at the inscription: my parents had given it to me on February 10, 1960. Dad died on February 10, 2001, precisely forty-one years later.

When we returned in December from Oregon to State College, what with bad winter weather and landing in Washington, D.C., we did not reach our quiet, suburban street until after three in the morning. Yellow ribbons on the trees, American flags, and signs festooning the street welcomed us home. But because of the hour, the neighbors who had planned to be outside to greet us had almost all gone to bed, and the only people there to meet us were reporters, including a half-dozen with television cameras. We posed for a few photos but asked to be excused from making statements, as it was very late and we were exhausted. They understood and left us alone to our tasks of release, recovery, and reconstruction.

The three tasks were interwoven. To achieve true release from my 253 days in a Russian prison, I had to recover my mental and physical health, and in order to move toward a new life I had to reconstruct and understand what had befallen me in Russia, and why. The writing of this book has served all those tasks, though the tasks themselves remain, and I suspect they will always be with me.

My first realization, one that had begun to dawn on me early during my imprisonment, was that after this event our lives would never be the same as before. The departure from the old norm brings with it great promise as well as heartbreak; in many ways my experience has made me more optimistic about my future than I used to be. I know I'm not going to waste my time being bitter, and I choose to turn the other cheek to the wickedness done to me. Today I'm excited, happy to be alive, very appreciative of each day, of the world around me, and of having the rest of my life to live outside prison walls, outside of Russia, and within a society that I now believe in more passionately than I ever did before.

But a good deal of my future has had to be put on hold while I set to investigating what happened to me in Russia, and why. The reader has already reaped some of the results of my labors — for instance, the reports in earlier chapters of the wonderful extent of the assistance given to my wife and children by my parents and sister and her family, by my other relatives, friends, former colleagues, and strangers. I will forever remain grateful to those who helped us and will continue to owe debts of more than gratitude to them; there is no doubt in my mind that without their assistance I would still be imprisoned in Russia.

Some of what I found out was amusing. For example, it was a laugh to learn how seriously the State Department took my every communication — about food. During the first weeks of my incarceration, Brad Johnson had worked with Cheri and me to establish a means of delivering fresh fruits, vegetables, and other personal items that were not being provided by Lefortovo. It took almost three months before Cheri's money could be properly received at the embassy and produce could be bought and delivered. Brad had asked me to make a list of what I would most want. Tired of the bland prison diet, I was ravenous for something fresh. To make my list simpler, I decided to put my requirements into three categories: (1) fruits and vegetables; (2) other food; and (3) personal items such as toothpaste, shampoo, socks, books, paper, and so on. I put a lot of thought into this list and rearranged it several times. There were approximately forty items, and to avoid Brad's having to write a long list each time he came to visit me, I developed a simplified code so that, after the first time he copied the list, he would then only have to write down a letter and number. As an example, I could tell him that I wanted F5, F7, P4, P11, and M9 and he would know that I wanted kiwi fruit, tomatoes, white chocolate, smoked fish, and shampoo. Brad was enthusiastic about this simplified process, and used it readily.

Only after my release did I learn that this "code system" had been viewed by Brad and the people following my case at the State Department as an attempt to provide coded messages of a far more important kind. A team of government analysts, investigators, and cryptanalysts were assigned to try to reconstruct the code they thought was hidden in my messages. My friends were canvassed for hidden meanings behind my requests for kiwi fruit, shampoo, and tomatoes; my home computer, books, and files were carefully searched for code words and encryption schemes that I might have developed. Powerful computers were used to see what hidden meanings there were in fresh garlic, apples, dark chocolate, and aftershave. Several weeks of concerted effort by the analysts led nowhere before the team threw up its collective hands and decided that I was either too smart for them or that I had simply been hungry and that there was no greatly significant code. Now it can be told: there was no code.

More seriously, since returning home I have found out many things about the behind-the-scenes dealings that precipitated my arrest, and the later dealings having to do with the timing and the manner of my release. I should mention in this connection that the State Department has since my release conducted its own investigation of its handling of my case and has revised some of its procedures based on that review. I believe that these new procedures have enabled them to be more helpful to the family of Jack Tobin, the Fulbright scholar imprisoned in Russia during 2001 for alleged drug possession. Also, the Office of Naval Research is currently conducting a review of the actions of Penn State ARL in reference to my case.

I am still compiling a third type of information. It is highly technical but extremely relevant: the evidence that every bit of information about the technologies that I was accused by the Russians of stealing could have been obtained from open-source material available in the United States.

After I returned home, I started searching for various references toward which I had been pointed by Russian scientists, such as the idea that the Russians had begun their HRG work after purchasing a license from AeroJet General in 1962. I have found that patent here, and also many others that have a direct relationship to the HRG that I was accused of stealing from Russia. I found a copy of the 1975 book that the "mad scientist" wanted me to obtain; the Russians might have based part of the Shkval's machinery on it. The most revealing work I've uncovered is a 1964 collection of research papers, *Underwater Missile Propulsion,* which includes a significant amount of work done at Penn State ARL. Had ARL forgotten they had done such work when they sent me to Russia to obtain reports on HRG from Bauman and Babkin? To make a long story a bit shorter: I've proved that the "secret" of the Shkval was developed right here in the United States of America and in other Western countries in the 1950s and 1960s — and *then* it became a priority target of Soviet agents who were collecting technical information from us and was brought back to the U.S.S.R. for further development. And develop it they did. But there are no "secrets" in the Babkin reports that are not also contained in a variety of ancient materials that I've unearthed in the U.S. (A selection of the relevant titles is included on my website, www.edmondpope.com.) The twisted reality behind my arrest and trial was that the big secret the Russians accused me of stealing was no secret.

Many things have happened in Russia since the time of my release, and some shed additional light on my case and the current state of affairs in Russia. The most directly connected to me involve the airing of a Russian television "documentary" about the case just days after my release and the publication in Russia, in the spring of 2001, of the book *The Spy Adventures of Pope in Russia.*

Since these are generally unavailable in the U.S., I have had them translated and can comment on their contents. Both are windows into the soul of the FSB and the ways that the Russian people are being led to think about themselves, their FSB, the United States, and the dangers to Russia from the West.

The "documentary" aired on the Russian state channel ORT on December 28, 2000. Entitled "The Secret of the Shkval Missile: The Edmond Pope Case," it purported to objectively ask and answer the questions of whether I had really been a spy, and why I had been pardoned. The program's bias was obvious from the first moments when, after showing pictures of the Shkval being tested, the Presenter says, "The man who devoted twenty-five years of his life to foreign intelligence knew very well the Shkval's unique features."

"We will reveal to you," the Presenter continues, "the main secret: the real and main reason for the U.S. interest in the Shkval missile." The explanation is something out of science fiction:

> U.S. analysts put forward a new military doctrine. In addition to developing the national defense system, U.S. scientists proposed that ballistic missile silos should be situated in the ocean.* And a lot of laborious work has been done in this area in the past four years. Just imagine, the very fact that Russia has this underwater missile [Shkval] wipes out the entire new U.S. military doctrine.

The Presenter walks with an unidentified FSB agent toward a hotel called the Ismailovskaya, near the great flea market of the same name that I liked to frequent. I stayed at the hotel in 1994.

PRESENTER: What were his rooms like?
FSB MAN: They were mainly single rooms.
PRESENTER: No luxury rooms?
FSB MAN: No luxury rooms. Just single rooms. Perhaps for financial reasons.
PRESENTER: Was any secret filming done in this hotel?

*There is absolutely no need for such silos under the ocean if we have — as we do — nuclear submarines capable of firing missiles.

FSB MAN: No, we did not do any secret filming in this hotel. It was just the beginning, you see, and Edmond Pope was not showing any particular interest in Shkval.

The FSB claims in this film to have become certain that I was focusing on the Shkval as early as 1996 and to have tracked me thereafter. Shakhidzhanov, the Region director, appears on screen to say, "We got the impression that we were having to give away [to Pope] a great deal of valuable information cheaply. It was quite unacceptable, and so it immediately became obvious that we had no commercial interest in this."

After this bit, the "documentary" excerpts the most damaging portions of the peephole video done at the Sayani Hotel and takes shots at my partner, Bolshov: "For Pope, Babkin and Bolshov weren't a bad team: they both perfectly understood their commitments and carried them out well." The Presenter then introduces "the last hero," Kiely, describing him as "the brain center of the trio . . . the one who . . . decided precisely which secret information had to be obtained." They show the agents pushing into the room where we four are sitting, and some of the interrogation that followed; commentary on the scene is provided by Aleksandr Shabunin, identified as deputy head of the investigation directorate of the FSB:

> His [Pope's] line of behavior, in the event of failure, had been thought out in advance. Of course, he couldn't deny any objective facts — he's quite an intelligent man. He confessed to what had been filmed and recorded by tape, but he only confessed from the point of view of his commercial work, which he presented as his main line. He saw the investigation as a mandatory procedure — "I got caught; go ahead, you do your work and I'll do mine."

One snippet of the interrogation in the Sayani Hotel room reports me as having initially said that as far as I knew, I had received and carried no classified documents. But then there was a question about a specific document. In English, on the hidden-camera videotape, I describe that document as dealing with a "sensitive technology"; but the Russian translation of that statement by the FSB interpreter in the room — which is all that the majority of the audience for the "documentary" would have been able to understand — makes me sound as though I have said that the document dealt with a "classified technology," in other words, that I have just contradicted myself and admitted that I knew the documents were full of classified matters.

This must have been one of the moments that initially convinced the FSB that I was a spy. But the supposition was based on a mistranslation by their own employee!

The "documentary" contains unnecessary lies and unsubstantiated accusations. In the first category is the allegation that I refused to share a cell with "two Negroes." Three Africans in the cells became my friends, and I am still in touch with them by mail. In the second category is the Presenter's remark that as a spy, Pope "did not succeed in fully accomplishing his mission," a statement that undercuts the prosecution's case against me in court, which was based primarily on the Babkin papers having been sent to the U.S. If those papers were so important, and if the FSB had had me under surveillance since 1996, why had they allowed such valuable documents to be mailed and e-mailed out of Russia and into my hands?

Significant also are the closing remarks of the "documentary," which avow that while Pope would "never be allowed to enter Russia again, the work [of U.S. espionage] will be continued," and that there might already be in Russia "another American businessman . . . who has come . . . to find out the secret of the Shkval rocket."

Published in Moscow in the spring of 2001, *The Spy Adventures of Pope in Russia* was written by Vasily Stavitsky, whom the book identifies as "a professional counter-espionage expert and writer," but who my contacts in the U.S. intelligence and business communities say is an agent of the FSB. His photo is on the back cover of the book, and I recognized him as one of the men who I had seen hanging around the halls in Lefortovo. His book is a more detailed reflection of the mind of the FSB and what that organization perceives as the most successful tactics for convincing the Russian population that they are in danger from foreign spies and need a strong FSB to protect them. These themes are sounded, right from the foreword:

> In recent years the special services of a variety of foreign powers have used legal intelligence methods in Russia to obtain secret information regarding modern industry and military technology. In the course of doing this work, they seek to undermine the national interests of Russia. Thus, under cover of one American firm, a professional spy, and officer of the Intelligence Directorate of the Ministry of Defense of the USA,* Edmond Pope, using the trust of Russian scientists, gathered secret information regarding a high-speed torpedo known as the Shkval, owned by the Russian Navy.

The text begins with a purported scene in a restaurant on the fringe of Penn State, where a Russian scientist and I are supposedly meeting on September 14, 1998. Bristling with dialogue, the scene baffles the reader by means of its high-tech baloney and the tossing around of names like Lockheed-Martin, the Kurchatov Institute of Nuclear Energy, and NASA-Lugar,† and terms like glass-fluid optical elements, on-board rigging of space systems, and fuel composition for the second and third stages of RS-22 and RSM-52. It makes claims that the lay reader might plausibly believe, but that scientists and engineers would know to be false, such as that there is very little in the world literature about the supercavitation principle. Quite a lot of information has been published on that principle.

*We have been unable to translate this precisely; our best guess is that this refers to the Defense Intelligence Agency, or DIA.

†There is no such place as NASA-Lugar; there is a Nunn-Lugar funding program that provides assistance to the former Soviet Union.

The scene switches to "the trail of a crime," at the Bauman Institute on March 26, 2000, where a routine(!) search of Babkin's computer by the institute's security department reveals that he is "leaking" classified information to me. Incidentally, Babkin's actual name is used in this scene and then changed to an alias for the rest of the book, in order to "protect" him. The same is done for other scientists, for Bolshov, and for many other Russians in the story, including such people as translator Alyosha and my first FSB attorney, Avdeev, both of whom become females for the purpose of the book. Here, too, the scene contains a lot of high-tech baloney, such as the serial numbers of the computers and printers and operating systems and e-mail servers used — as though such specificity guaranteed the accuracy of the reportage.

The main accusation of the early sections of the book is that I was operating as a spy under commercial cover, a subterfuge that had been made legitimate in the U.S. by a law passed in 1991. Such a law was passed, but it had nothing to do with my commercial ventures or any other employment. A Colonel Sidorov of the FSB suggests that I had been "retired" from the Navy for the sole purpose of enabling me to resurface as a private businessman, an arrangement that "allows the U.S. government to avoid culpability and conviction of espionage." A later chapter marvels at the "breadth, insolence and cynicism" of my efforts, and those of the U.S. in general, to "collect classified information on Russian soil."

In his own broad, insolent and cynical way, the author then compiles a list of documents that are so secret that he cannot reveal their contents, but does print their "opening and closing words only." The phrases he cites are evocative and cumulatively sinister:

> Stevens began the closed meeting by saying, "I ask that you lay out for me in the next hour, your program for your visit to Russia."
>
> "Well, the main goal is to obtain the fuel characteristics of the Shkval underwater missile," the intelligence chief pointed out. "Continue."

Another section:

> "Well, as far as a 'source' goes, Captain, you are satisfied with very little. Where is the recruitment report?"
>
> "This is one I would characterize, sir, by saying, 'More like a fool than an agent.'"

Another good mirror of the FSB mind is the set of one-word tags for individual scientists that the Russians accuse me of writing in my address book: "goat, fool, swindler, tubby, bumbler." I wrote no such tags in my address book or elsewhere, but such words are frequently used by Russians to characterize underlings who become tools of more powerful people.

A theme harped on repeatedly in the book is that I signed this or that interrogation protocol, as though by signing I agreed that the protocols contained truth and not falsehoods about their subject matter. That, as the reader of this book already knows, is a misrepresentation. Furthermore, some of the *Spy Adventure*'s reproductions of protocols contain outright lies, such as the author's (or the FSB's) assertion that I had met with a State Department official in the summer of 1999 who forbade me to develop a contract with Bauman. Curiously, though, the Stavitsky book also reproduces protocols that sustain my version of events and specifically refute the contentions of the prosecutors during my trial. The prosecutors charged that I had personally paid $30,000 to Babkin and Bauman; the Stavitsky book reports on page 59 a protocol which makes the point that — as I had insisted all along — Penn State ARL had made four wire transfer payments to Bauman for each of the first four reports, for a total of $28,000. The last payment, for the last report, was made by me to Bolshov, for transfer to Babkin.

The book flits back and forth between what purport to be verbatim reports of protocols and imagined scenes in America — for instance, one in which a Department of Defense official threatens to demote a rear admiral to midshipman if he does not send certain documents to Moscow within three days, and the admiral then threatens to replace a captain with a younger deputy and tattle on the captain's secret illnesses if he doesn't obtain the documents pronto. Such threats and motivation-by-blackmail are routine in Russia, but not in the U.S. The relatively few pages in the book devoted to my trial — it was, after all, a "closed" trial — do not mention such important aspects of the trial as the recantation by Babkin, the replacing of the prosecutor, the introduction of evidence supporting my contentions that the information I sought and obtained was from open sources, or that the Kiely protocol was disallowed.

Perhaps my favorite part of the book, simply because it made me laugh, was the description of an aspect of my arrest: "Nobody tried to run away, well understanding that to run in a race with strong young studs from the FSB was senseless."

Within Russia, many important changes have occurred in the wake of my case.

Immediately after my release, Putin attended the seventieth anniversary of the Chekists, a gala held at the Kremlin (not at Lubyanka) for the old KGB and its successor agencies. His presence and laudatory remarks there were widely circulated. To help celebrate the event, a memorial CD was made of Chekists fondly remembering Rudolf Abel,* and a move was begun to restore the KGB name to the security agency. More recently, Putin has appointed several more former St. Petersburg KGB men to high positions in his administration.

The Pardon Commission has been all but deprived of its function. Although in the year 2000 some 12,000 people were set free from prison, in the first six months of 2001, no Russians were pardoned. Commission chairman Pristavkin told reporters in June 2001 that the only person who had been pardoned since September of 2000

*Abel, convicted in the U.S., was exchanged for Francis Gary Powers in 1963.

was Ed Pope, although his group had recommended about 3,000 cases to Putin for pardons. According to another member of the commission, human rights activist Valery Borchov, the commission is "standing on the verge of destruction," because the Ministry of Justice wants to put the pardon process under its control, rather than have it function in effect as a check on its excesses.* Commission members have been unable to get President Putin to sit down with them to discuss the impasse. An anonymous but highly placed Kremlin source told the *Moscow Times* that "the Justice Ministry is guilty of trying to step between the president and the prisoner, which they have no right to do under the Constitution. The ministry simply doesn't want society to have a say in it [the pardon process]." Copies of internal Ministry of Justice memos obtained by the *Moscow Times* revealed that prison officials were being asked to crack down on the number of petitions for pardon sent out by prisoners.†

Another significant change: in June 2001, the Russian Academy of Sciences sent a directive to its member institutes and scientists ordering "constant control" over its scientists' contacts with foreigners in order to prevent espionage, including limitations on scientists' trips abroad and on participation in international conferences held within Russia by "researchers who have access to state secrets." According to the Associated Press, at least one research institution, the Institute for General Genetics, turned the directive into a policy memo ordering its scientists to report to their supervisors all contacts with foreigners. In an interview on *Echo Moskvy,* Duma deputy and human rights advocate Sergei Kovalyov called the directive "dangerous" because it was "quite in line with today's Kremlin policy," which he characterized as a throwback to the Soviet era.‡ Konstantin Preobrazhensky, a for-

*Having seen at first hand that the Ministry of Justice functioned as an arm of the FSB, I believe that this realignment is tantamount to placing the fox in perpetual charge of the chicken coop.

†Ana Uzelac, "Why Putin Is Not Signing Any Pardons," *Moscow Times,* June 1, 2001, and "Red Tape Leaves 3,000 in Prison Limbo," *Agence France-Presse,* June 28, 2001.

‡Anna Dolgov, "Academy to 'Control' Scientists," *Associated Press,* June 1, 2001.

mer lieutenant colonel in the KGB who had once been involved with the KGB's efforts to keep tabs on scientists, wrote recently that the actions of the Academy of Sciences presidium amounted to a "grave assault on the freedom to exchange ideas and information," an assault that would "lead to nothing less than the destruction of Russia's academic intelligentsia." He added that he knew personally that the majority of Academy members were now either outright agents of the FSB or had been so co-opted by the security services that they functioned as agents, reporting on their peers. "Every time an Academy member complies with the regulation," he wrote, "he or she merely opens up an opportunity for an alert [FSB] officer to justify his salary and position." The deeper problem:

> Peering out from behind today's FSB, one increasingly recognizes the familiar face of the KGB's Fifth Directorate, which was charged with combating domestic dissent. In all the countries of the former socialist camp — except Russia — the officers of such directorates have been dismissed from service. Here, they have remained and are now occupying leading positions.*

Elena Bonner, former Soviet dissident and now a resident of the West, finds the whole of the new Putin regime "permeated with lies" and characterizes my trial, and those of Nikitin, Pasko, and Moiseyev, as part of a pattern of "obviously trumped-up" cases that is of a piece with "the expansion of state control over the mass media," the lies put out about the sinking of the *Kursk,* and the shameful conduct of the war in Chechnya.† The common thread in all of these cases, adds another émigré, chess champion Gary Kasparov, is fear —

*Konstantin Preobrazkensky, "Academy of Spyences," *Moscow Times,* June 9, 2001.

†Elena Bonner, "The Remains of Totalitarianism," *New York Review of Books,* March 8, 2001. A recent audit by the Duma discovered, according to an authoritative source, that "all the money earmarked for rebuilding Chechnya had been embezzled." William E. Odom, "The End of Glasnost," *Wall Street Journal,* December 14, 2000. Odom is a former Army general and director of the National Security Agency.

fear among ordinary Russians that "their country is under attack from hostile forces. . . . Instead of beating the real hostile forces in Russia — corruption, ignorance and a bloated state — Mr. Putin has cleverly changed the rules of the game."*

As Bonner suggests, I was incredibly fortunate to get out of the clutches of the Russian FSB and judicial system. Valentin Moiseyev was convicted for the second time in August 2001. Two more people who are probably innocent of espionage but are still being held on such charges are Igor Sutyagin (researcher for the U.S.–Canada Institute) and Valentin Danilov (accused of spying for China). During the Sutyagin trial, the government gave to television and radio news programs what purported to be a tape of Sutyagin conferring by telephone with Captain Robert Brannon, the senior naval attaché at the embassy, and discussing in English what sort of armaments were aboard a Russian intelligence ship sent into the waters off the Balkans during the NATO campaign there; the FSB provided the news media with its Russian translation of the conversation. A few days later, what the *New York Times* called "a closer examination of the broadcast . . . revealed that the Russian translation [of the conversation in English] does not comport with the actual conversation. Most importantly, the central allegation made in the voice-over cannot be heard on the tape recording."† Why am I not surprised at such FSB tactics?

Grigory Pasko's second trial is in the works. Anatoly Babkin is now on trial ostensibly for being in cahoots with me. Pavel Astakhov is under investigation by the FSB for his supposed leaks to the press of information from within my closed trial. Young Jack Tobin, accused by the Russians of espionage, then convicted only of drug possession, was released from prison mainly because of the efforts of his family, his congressman, and the press. In many senses, I feel

*Gary Kasparov, "I Was Wrong about Putin," *Wall Street Journal*, January 9, 2001.

†Patrick E. Tyler, "That Russian Espionage Tape Was Not Quite All It Seemed," *New York Times*, March 30, 2001, and see also Tyler, "Russia's Spy Riposte: Film Catches Americans in the Act," *New York Times*, March 28, 2001.

like the last man to escape the awful fortress before the gates were slammed shut.

That impression was reinforced by a recent story in the *Washington Post.* In 1984, a KGB museum was opened in Moscow by Putin's childhood hero, Yuri Andropov, and in April of 2001, journalists from the *Post* visited it and found a lot of evidence to support the idea that there is a resurgence of Cold War attitudes in Russia. The guide, a former KGB officer, chided the Americans on the treatment of Russian citizen Pavel Borodin, who was then in a New York jail awaiting extradition to Switzerland on charges of corruption and embezzlement, and claiming to be too ill to travel. "In this country, that would never happen," the guide asserted. The journalists countered with a question about Ed Pope, who had been refused permission to see American doctors while in a Russian prison, despite his fear that his cancer might recur.

"He hasn't died yet," the guide retorted.*

Some friends of mine in the U.S. interpret the guide's remark as a threat: The FSB will kill you yet, Pope!

I take it as a reminder of how important my case was to the FSB in terms of its need to appear as though it is maintaining Russia's vigilance against foreign spies, and in terms of relentlessly trying to disprove anything negative (or embarrassingly truthful) that outsiders might have to say about the security services.

The many changes in the Russian Federation over the past year demonstrate that the Putin administration has been and will continue to be under the thumb of hard-liners who will not stop until they achieve full dictatorial, Soviet-style control over the populace of the Russian Federation. These hard-liners consist of reactionaries in

*Susan B. Glasser and Peter Baker, "The Cold War, Frozen in Time at the KGB Museum," *Washington Post,* April 2, 2001. Just days later, Borodin agreed to be extradited to Switzerland; within weeks, he was free, because Russian agencies refused to provide any evidence against him, and because his $2.9 million bail was paid by the Russian Ministry of Foreign Affairs. See John Tagliabue, "Extradited Russian Suddenly Free After 3-Year Inquiry," *New York Times,* April 13, 2001.

the military, people from the FSB and other police and intelligence services, and the Mafia. In combination, they run the government. Fomenting exaggerated suspicion of the West through spy-mania is a featured aspect of their attempt to reassert total dominance of the population.

There is also some evidence that the Putin administration is implementing a strategy to reconstitute the old Soviet Union by means of "technical-military cooperation agreements" with former Soviet states such as Belarus and Kazakhstan that will increasingly bring those states under the Russian Federation's sphere of influence, and to reestablish its arms-supplier relationship to Iran, Iraq, India, North Korea, and Cuba.*

I will never return to Russia, but I remain a Russophile, a lover of the country's art and music, and a great admirer of its scientists and technologists. Their energy, their level of intelligence and knowledge, their unusual approaches to problems, have been remarkable. But the Russian technological institutes are like beautiful flowers in a garden choked by the weeds of the military past. The death of some of the institutes will be caused by these weeds, and other institutes must be allowed to wither and die before the ground is retilled and the remaining flowers reach their peak of blooming. If only the focus of these institutes, and of Russia's scientists and technologists, could be redirected toward nonmilitary technologies, their achievements would positively further the world.

That goal may take a long time to achieve, because the key watchword of Russian life is still "survival," and it is very difficult to change for the better when you are worried about how you will earn enough money to pay for your next meal.

Unless and until that goal has come nearer, not only for the technological institutes but for all of Russia, the peoples and the governments of the United States and of the other Western democracies

*"Military Deals 'Way of Influencing World Power Shifts,' Says Putin," *Agence France-Presse*, June 11, 2001.

must remain very wary of Russia, and especially of its ruling elite. The Putin government cannot be trusted, and neither can individual Russians, in most instances. The Putin administration is racing Russia back to its totalitarian past at tremendous speed, embracing the tactics the Communist governments used to control the populace, but with a new twist. Since many of the ruling oligarchs are making lots of money from the new, supposedly capitalist and free-market economy, the rulers are opposed to the central economic control that was a feature of Communist governments; so Putin and his minions are combining the worst aspects of Communism with the worst aspects of Fascism.

Had it not been for public pressure on President Putin, the hard-liners in the Kremlin would have prevailed and my release might never have taken place. That pressure came in part from the American government, but it must also be said that public pressure played the most significant role on the American side, as well. The willingness of those at high levels in the Clinton administration to press for action from the Putin government was strengthened by the congressional resolutions that were themselves the focus of public pressure, and those officials' readiness to work hard for my release was also spurred by the attention that the media gave to my case.

It is good to be home.

Beyond remaining grateful for all that was done on my behalf, I am proud that it could be done, and was so readily done, in my country, and by my family, my neighbors, old shipmates, military and ex-military personnel, friends, friends of friends, and total strangers. The large number of people who expended time, energy, and ingenuity working toward my release, and the success of their efforts to bring my plight to the attention of government officials and to move those officials to action, is to my mind a grand expression of the American spirit and a vindication of the American democratic system.

In my recent travels around the U.S., I have frequently met people I had not known before, and have been told by them that during my imprisonment they had been praying for me regularly in their church groups or veterans' groups. Such actions reflect a great strength of our society that Russians do not understand and that astounds them: that in the United States, ordinary people do things that contribute to the well-being of their fellow man on their own, without being directed or compelled to do so by the government. My experience has shown me, and I hope that it helps others understand, three things of absolute importance to our way of life: freedom of the press; the basic idea that in the eyes of the law we are all innocent until proven guilty; and the recognition by the government of the need to constantly uphold the rights of individual citizens while considering the needs, roles, procedures, and foreign policy of our nation.

Acknowledgments

I want to first thank my immediate and extended family, especially Cheri and my two sons, Brett and Dustin, my mother, Elizabeth, and my sister, Brenda. In many ways, they suffered more than I did from the incident of my imprisonment. They all stood up and did everything they could, from all corners of the country, to protect me and ultimately to secure my release.

Another group I want to highlight includes Congressmen John Peterson and Curt Weldon, both of Pennsylvania, and Congressman Greg Walden of Oregon. John Peterson and his staff were totally dedicated and focused on our situation from start to finish. Many other members of Congress were helpful and followed the lead of John Peterson in eventually securing my release. In my mind, John Peterson is a true American hero and exemplifies the qualities we all seek in a congressional representative.

There are many individuals who stood out from the hundreds who offered Cheri support. Primary among them is my friend and partner Keith McClellan; others include Tom Brooks, Ted Daywalt, Dave McMunn, Dave and Joan Moss, Brad and Jay Mooney, Bud Moore, Gary Hade, Jerry and Patsy Ballinger, Mike and Judy Keating, Shap Shapiro, and Sid Wood. My only misgiving in mentioning these individuals is that I know I am leaving out many others who deserve special recognition, some for their own protection.

Many organizations, such as the Naval Intelligence Professionals, the Veterans of Foreign Wars, and the American Legion were supportive as groups as well as on an individual basis. Many church groups and individual churches followed our case and provided prayer and other forms of support on a regular basis.

I want to thank Tom Shachtman for his literary talents as well as his true friendship. Without him, the book would certainly have had a different, probably far less compelling, tone. I have also been totally delighted with the support and interest in our situation from the many people at Little, Brown and Company, Time Warner, and the William Morris Agency, particularly Norm Brokaw's personal attention.

Finally, I want to acknowledge and thank the hundreds of thousands of people all over America and in other countries who prayed for, followed, and otherwise were sympathetic to our plight. You can never fully appreciate the blessings bestowed upon us until you find yourself involved in a difficult situation. The hundreds of thousands of people we have never met, yet who did support us, are part of this story, and I hope that they will share our newfound appreciation in a free, democratic society.

INDEX

Index

Index

LaVergne, TN USA
16 May 2010
182868LV00002B/24/P
9 780316 348737